The Peril and Promise of Medical Technology

New International Studies in Applied Ethics

VOLUME 8

EDITED BY

Professor R. John Elford and Professor Simon Robinson,
Leeds Metropolitan University

PETER LANG

Oxford · Bern · Berlin · Bruxelles · Frankfurt am Main · New York · Wien

The Peril and Promise
of Medical Technology

D. Gareth Jones

PETER LANG

Oxford · Bern · Berlin · Bruxelles · Frankfurt am Main · New York · Wien

Bibliographic information published by Die Deutsche Nationalbibliothek.
Die Deutsche Nationalbibliothek lists this publication in the Deutsche
Nationalbibliografie; detailed bibliographic data is available on the Internet
at http://dnb.d-nb.de.

A catalogue record for this book is available from the British Library.

Library of Congress Control Number: 2013946126

ISSN 1663-0033
ISBN 978-3-0343-0775-8

© Peter Lang AG, International Academic Publishers, Bern 2013
Hochfeldstrasse 32, CH-3012 Bern, Switzerland
info@peterlang.com, www.peterlang.com, www.peterlang.net

This publication has been peer reviewed.

Printed in Germany

Contents

Foreword by General Editor

This book makes a major contribution to the debate about ethics and the use of technology in medicine. It addresses a Christian Ethics perspective, critically engaging the theologians who are largely involved in defence of faith positions. It also provides a contribution to the wider philosophical debate in this area. The reason why it is able to do both so well is because Gareth Jones' focus is on the scientific, technological and caring narratives at the centre of the debate. Hence, with each aspect of the debate he is able to both guide us unerringly through the often complex material and take us to the human heart of ethical decision making. What he finds there is not certainty, but many different value narratives, even within the Christian religion. Key to ethical decision making is critical engagement of those narratives, taking us beyond a simplistic application of natural law. This enables Jones to look in depth at issues such as the technological enhancement of morality, ageing and identity, and humankind's relationship to technology. His conclusion takes us away from the hyperbole and 'slippery slopes' so often evident in the debate in this area, to a hope that is based in rigorous critical dialogue and reflective and responsible scientific practice.

Simon Robinson
July 2013

Preface

This book follows on from a number of previous ones, all of which have dealt in their various ways with the interplay of biomedical technology and Christian faith. Coming back once more to this topic area demonstrates a number of things. There are no definitive answers that will provide lasting guidance or assured answers. The reason for this may lie in the nature of the Christian faith or possibly reflect my own approach to these issues. However, I do not think this is an adequate response. The issues themselves are changing, the technology is always on the move, and therefore the debates are undergoing modification from one year to the next. Of course, one could get around this relentless change by simply asserting that this or that technological intrusion is to be opposed; if we lived in an ideal world they would be prohibited, and even if that is not the world in which we live, we ourselves should be resolutely opposed to them. It is not difficult to find groups who adopt exactly this position, with the result that books emanating from them will adopt precisely the same viewpoint today as they would have done in the 1980s. Generally, they will find little need to delve into the pros and cons of new technological developments, since their stance will be known in advance.

These comments should not be interpreted to suggest that the debate is going around in circles and getting nowhere. The parameters are shifting, and the precision and even the nature of the questions is changing. In the Preface to my 2005 book *Designers of the Future* I reflected on these matters in much the same way. I wrote: 'the pace of scientific developments has necessitated a re-examination of what I wrote just a few years ago both to bring it up-to-date and also to reassess it. This does not mean I have made substantial changes to stances I held in the 1990s, but a fair degree of nuancing has inevitably taken place.'[1]

---

[1]   D. Gareth Jones, *Designers of the Future: Who Should Make the Decisions?* (Oxford: Monarch, 2005), 8.

While I do not wish to concentrate excessively on beginning-of-life issues, there is no escape from them – at least for me. This is because they appear repeatedly in one guise or another. They are foundational for Christian thinking on biomedical issues and probably for many other worldviews as well. They are never far away, and so even though one may have a frustrating sense of déjà vu, they inevitably keep raising their heads and refuse to go away.

As I have stated previously, I come to this material as a scientist with a strong interest in bioethics and functioning within a Christian framework. I claim no theological expertise, beyond a commitment to a Christian ethos and a deep interest in the way in which Christians might best approach the data and interpretations coming out of biomedical endeavours together with the theorizing surrounding these.[2] I am also committed to providing 'ordinary' Christians with practical assistance when confronted by the seemingly intractable dilemmas thrown up by contemporary biomedicine. This is because we are dealing with intensely practical and on occasion personal issues, far removed from the world of academia, whether theological or philosophical. One consequence of this approach is that I aim to confront highly contentious options, and emerge with what I argue are options amenable to Christians and those of like mind. I refuse to leave the discussion at the level of general principles. I would much prefer to test possibilities, as a way of providing helpful steps forward and also formulating hypotheses that can be tested and, if found wanting, discarded.

Originally I thought of using the term 'human nature' in the title, but on further reflection dropped the term since I was reticent to try and define it. Its vagueness is unhelpful, although there may be well-defined and perhaps fixed limits to central human characteristics, such as thinking, feeling, acting, worshipping and loving. Can these characteristics be moulded and can they be radically altered by the activities of human beings themselves, and in so doing will the essence of human beings be changed?

---

2   This is not an account of bioethical issues that is distinctly theological. I am not
     equipped to provide such an account. I leave accounts of that nature to writers such
     as Allen Verhey and Neil Messer, both of whose writings I find very helpful.

Will they become something else, whether Frankenstein monsters, cyborgian creatures or post-persons? In contemporary discourse any possibilities along these lines are technology-driven, with biomedical technology at the forefront of all that might be good or bad.

It requires little imagination to realize that this is a realm of intense relevance to Christian theology. No longer are these possibilities of theoretical interest alone; they are of practical significance. The way in which people act, the way in which they think, their hopes and prospects are all bound in to the new world in which we find ourselves in Western societies. This is true for those with a faith-based perspective, as much as for those without this. The cultures in which we live are being transformed by the shift in expectations brought about by technologies that influence how long we live, the quality of the health resources available to us, our dependence upon extensive arrays of drugs and medicines, the nutritional status of our diets, our participation in fitness regimes and even our dependence upon appearance medicine clinics. No one of these is entirely positive or negative in the role it plays, but together they dramatically alter human expectations and human activities.

In looking at these challenges, I accept that medical technology is and will remain a dominant force in human life. From this there is no escape as long as the resources are available to continue to support high levels of technology. Throughout I shall come back to the theme of hope, a concept I see as a central Christian affirmation. But it is a notion that appears to be challenged by many of the allures held out by medical technology, with its prospect of alleviating all human suffering and angst, and of providing a pathway to a new world in which hope stems from the technology itself. The idealism inherent within movements like transhumanism and moral bioenhancement is poignantly suggestive of this end result.

I am no idealistic promoter of everything brought into being by the purveyors of the latest technological gismo. But I am committed to a realistic assessment of the possibilities, rather than concentration on the extremes. It is easy to advocate for or against 'science fiction scenarios', albeit dressed up as the darlings of current debate. Gene manipulation to produce smart, tall, fair-haired, compliant individuals entices some and appalls others, and lends itself to being the subject of intense debate.

Fascinating as this may be, it is relatively easy to deride and reject such a scenario, but far more difficult to decide on the ethical (and theological) validity of utilizing screening procedures in the early stages of pregnancy to test for Down syndrome or cystic fibrosis.

I shall argue that this distinction between the extreme and the realistic is a core consideration for any approach to the inroads of biomedical technology. The reason is that concentration on the extremes protects us from serious analysis of the realistic, which is where science and theology meet. I shall insist that we never stray too far from the realistic and the imminent, since this is where ordinary people encounter technological imperatives and look for theological guidance. If we fail at this point, no amount of speculation about what may befall humanity in 2060 or the twenty second century will have much relevance.

Another core consideration arises in the face of the technological-centered control now in the hands of human beings. What criteria do we possess to try and ensure that technology is used to help people rather than threaten them? Science can be cast as a welcome addition to the armamentarium humans possess to quell disease and deformity, but by the same token it may also push accepted boundaries and threaten human conceptions. I shall look for theological guidance in these contested areas, but it will always be informed by scientific analysis and insights.

This, in turn, brings us face to face with a further thread running through this discussion, and this is the nature of the theological approach I am adopting. As in previous writings I look as far as possible to biblical directives, although I am intensely aware that these do not provide anything remotely resembling neatly packaged answers. It is shortsighted to expect such guidance, and yet this is a statement frequently challenged in some quarters. It is all too easy to assert that a particular position is 'biblical', even when the evidence found in Scripture is sparse, and hermeneutic issues are legion.

The material here has been given in various forums, including Christians in Science and the Faraday Institute in the UK, American Scientific Affiliation in the USA, ISCAST in Australia, as well as in a variety of anatomical settings in the USA and China. It also builds on previous books, especially

Valuing People: Human Value in a World of Medical Technology,[3] *Designers of the Future: Who Should Make the Decisions?,*[4] as well as in chapters of two edited books with John Elford, *A Tangled Web: Medicine and Theology in Dialogue* and *A Glass Darkly: Medicine and Theology in Further Dialogue.*[5]

As in recent years I am extremely grateful to Maja Whitaker for her insightful comments and suggestions on all aspects of the book, which was finally put together after I had relinquished the reins of university administration.

D. Gareth Jones
Dunedin
June 2013

3   D. Gareth Jones, *Valuing People: Human Value in a World of Medical Technology* (Carlisle: Paternoster Press, 1999).

4   Jones, *Designers of the Future.*

5   R. John Elford and D. Gareth Jones, eds, *A Tangled Web: Medicine and Theology in Dialogue* (Bern: Peter Lang, 2009); D. Gareth Jones and R. John Elford, *A Glass Darkly: Medicine and Theology in Further Dialogue* (Bern: Peter Lang, 2010).

Challenges of modern medicine

The disconcerting world of modern medicine

In an introductory chapter to *A Glass Darkly*, I set the scene by asking whether medicine is out of control.[1] The reason for commencing that book of essays on the relationship between medicine and theology in this particular manner was that the character of medicine appears to be changing. Traditionally medicine has been regarded as a healing profession, in which as far as possible the ill have been restored to good health, and have been cared for and comforted even when cure has proved impossible. There has been a pastoral dimension to medicine that has fitted alongside that provided by other caring professionals. The increasing powers of medicine brought about by increased understanding of the workings of the human body need not upset this equilibrium, since the end result may simply be a welcome increase in the ability to cure a greater range of illnesses. However, has this increase in curative ability actually upset the equilibrium between cure and care? On the surface this need not be the case. Why should greater efficiency lead to a diminution in empathy and why should it pose a threat to Christian conceptions of the medical enterprise?

If there are threats they arise from changed conceptions of medicine itself and from changes to the goals of medicine. Much is made in some

---

1   D. Gareth Jones, 'The Biomedical Technologies: Prospects and Challenges', in D. Gareth Jones and R. John Elford, eds, *A Glass Darkly: Medicine and Theology in Further Dialogue* (Bern: Peter Lang, 2010), 9–32.

quarters of the Hippocratic tradition in medicine,[2] with its emphasis on the doctor's duty of care to heal the sick and a refusal to take life. For Wyatt this led to trust, philanthropy and separation between healing and harming.[3] While this alone may not be as powerful as sometimes asserted, since it was radically extended by Christian attitudes, it draws a line in the sand that is worth remembering when confronted by ever-increasing technological and clinical abilities. This is because contemporary challenges stem from the power of these capabilities alongside dramatic changes in the world-views of many in society. Consequently, 'the Christian drivers that led to the establishment of hospitals and overcame social deprivation have been replaced by a secular humanistic worldview intent on lauding biological quality and longevity at the expense of care for the disadvantaged and disabled'.[4] For his part Wyatt puts down the erosion of the 'Hippocratic-Christian consensus' to reductionism, technology, and consumerism.[5]

The reductionism referred to by Wyatt alludes to the trend to view humans as little more than physical machines to be understood merely as biochemical, physiological and molecular entities. When taken too far the result is that people are no longer understood within a web of social relationships and cultural norms. This is the foremost of the challenges we shall encounter in the coming pages, a challenge emanating from the increasing power of technology accompanied by pretensions that tend to overinflate this power. Very readily what can be done to manipulate brains, livers and limbs is viewed as a means of manipulating and transforming the individuals themselves. Rather than being content with healing and caring for these people, the manipulations become an end in themselves. People can be transformed into new beings on account of their manipu-lated brains, livers or limbs. Indeed no longer is it sufficient to allow people to continue with their original brains, livers and limbs; perhaps people

2   Nigel M. de S. Cameron, *The New Medicine: The Revolution in Technology and Ethics* (London: Hodder & Stoughton, 1991); John Wyatt, *Matters of Life and Death*, 2nd edn (Nottingham: Inter-Varsity Press, 2009).
3   Wyatt, *Matters of Life and Death*.
4   Jones, 'The Biomedical Technologies', 9.
5   Wyatt, *Matters of Life and Death*.

en masse should be given new organs and appendages, improvements on those with which they were born. These reflections, speculative as they are, point to some of the prospects held out by modern medicine, with grandiose and probably unrealistic expectations. The transition from helpful application of technological capabilities to transformative vistas about the nature of human existence is thought to be eminently achievable in a climate of technological dependence.

It is not surprising that many within a Christian context abhor what is viewed as the downward moral spiral brought about by the pretensions of clinicians and scientists. After all, theologians and theological ethicists would never go down such paths; they are far more cautious and are acutely aware of the ill-fated hubris represented by such endeavours. While this is an oversimplification, it makes a point. The ascendance of technology appears to pose threats to Christians, let alone to those of other religious traditions, not because of the technology itself but on account of the messages that accompany it. Technology, rather than being a useful tool, comes to be seen as a source of meaning by giving direction to human aspirations. This is the nature of the science-faith divide in the biomedical world and it lies at the core of the disconcerting messages that many take from developments within contemporary medicine.

The underlying issue appears to be whether medicine is intruding into the divine centre of human existence, especially at the beginning and end of individual existence. Individuals exist who would not have existed in the absence of medical intervention. Implicit within this simple statement is the feeling that humans can now decide on existence and non-existence in realms that until recently were completely outside their control. It seems as though chance has been replaced by choice, with the chance subtly being read as divine in origin, and the choice as emanating from human prowess. Once any interpretation along these lines is accepted, perceived conflict emerges: where once God ruled supreme, humans now routinely intervene. It is easy to gain the impression that humans have all the abilities necessary to displace God. Some scientists boldly make this claim, and even make derogatory comments about theology and theologians on the way. According to Dawkins 'theologians have nothing worthwhile to say about anything'.[6] Quite apart from these crass interjections, sentiments about the ever-increasing power of medical technology are rife with theological overtones. One particular Christian riposte is to contend that medicine is indeed becoming too powerful, since human beings cannot be trusted with such awesome power. Perhaps we should admit that the critics who view these developments as deeply flawed might indeed be correct; perhaps human beings are trespassing in treacherous, even forbidden, territory.

A recurring temptation for Christians is to strive to maintain the ethical and legislative *status quo* on the premise that the *status quo* represents a more acceptable position than anything that might emerge from current discussions. This is especially the case where issues touching on human wellbeing are concerned. This is an understandable response when confronted by the unknown, and yet it fails to address the central issue, namely, that the technology itself brings with it benefits. It is the way in which the technology is used and the aspirations held out for it that may be problematic. By seeking to maintain the *status quo*, no attempt is being made to confront these philosophical, social and even theological challenges. Any Christian response that bypasses these issues is unsurprisingly seen as being luddite; it will not stand up to the test of time, since much of the technology will one day be accepted – even by Christians – and the deeper challenges will have gone unanswered.

For me concern over the burgeoning of human abilities in these areas of tension elicits two immediate rejoinders. The first is that throughout history humans have had immense power over who is brought into existence. It is true this may have been confined to whether a new life is or is not brought into existence, with little consideration given to the nature of that life. But it would be unwise to downplay this power, even though it has been 'natural.' The second is that, as already mentioned, to designate choice as human-driven and chance as the domain of God or the divine is highly questionable. This is reminiscent of a God-of-the-gaps approach, where God is confined to the realm of chance. When expressed in these terms, the theological absurdity of the dichotomy becomes evident. There

6   Richard Dawkins, *The God Delusion* (London: Bantam Press, 2006), 57.

is no hint in this formulation that God is sovereign over the whole of life; neither does it acknowledge the God-bestowed creativity of human beings as expressed in the creation account in Scripture. Examples in Genesis include 1:28 where humans are to 'reign over the earth and govern it'; in 2:15 humans are to 'tend and watch over the garden'; while in 2:19 Adam is given the task of naming the animals. While a great deal of effort is required to decide precisely what each of these commands entails, the role of human beings in governing and overseeing numerous facets of the physical world shines through. There is no hint here that human beings are to sit back and let God, or nature, do all the work. Human beings are implicitly involved in ensuring that the physical world, including human life, functions as well as possible. There can be no abrogation of responsibility to a higher authority, although from a Christian perspective the ideal is working in tandem with directives provided by this higher authority.

However, there are deeper concerns, since implicit within the power of technology is the prospect that it could be used to change people in far more radical ways than has been the case until recent years – extending life to well beyond 100 years for most people in society, extending eye sight or hearing beyond anything contemplated let alone experienced up to now, increasing the creative powers of whole populations, or enabling people to communicate with one another using implanted brain probes. This is where human control becomes far more evident, although even here there are no foregone conclusions. The end-result of such control may be to improve the lives of people, along the same lines as conventional medical treatment has aimed to do for centuries. Over against this it could be used to improve people in a far more profound sense, namely, that they behave in far more moral ways than prior to technological intervention. This is a challenge for those, including Christians, working in the biomedical sciences, since their efforts are so often directed towards increasing an understanding of how the human body works and therefore how it can be made to work better. Is this effort legitimate if it enables people to live longer and to live relatively disease-free lives, to tackle severe genetic problems before birth, to have babies later in life than previously possible, or to think smarter and stay awake longer?

It is prospects of this nature that have proved deeply disconcerting for many people, including – but not confined to – those with religious perspectives. If humans are capable of transforming the means of functioning of ordinary people so that they become substantially different from what they would otherwise have been, have humans become creators – in a limited sense at least? The boundary between choice and chance has become exceedingly murky, as has that between therapy and enhancement, treatment and improvement. The former is associated with the routine regenerative processes of the body, whereas the latter appears to have superseded these. Once again though it would be unhelpful to jump to precipitate conclusions, since there are ways of improving regenerative processes in contemporary medical treatment, where the aim is to restore an individual's body to a reasonable functioning state and no more than this.

For Christians these possibilities also bring with them an array of pastoral implications, since they give the impression of making God more distant from their everyday experiences. There can be no escape from the question of where God enters the picture. One is forced to ask what role prayer plays when confronted by illness and the apparent effectiveness of drugs, surgery or brain implants. Even more troubling, are transformative modifications that may leave the individual more of an extrovert than before the technological intrusion, a gambler rather than their risk-averse self, or religious rather than someone who had previously never had a religious thought.

Through a glass darkly: ethical and theological uncertainty

Those of us in science, and especially in biomedical science, are caught in the crossfire. I am part of this inexorable onward drive of medical technology. My general attitude towards it is cautiously positive. In this I stand in stark contrast to a number of other Christian commentators who give the impression of being far more poignantly aware of the dangers that lurk around the corner.[7]

From my perspective we do not have the luxury of sitting back and deciding where we do or do not go. We have no option. This is not to say that we utilize every technology available under all conditions; in practice, this does not happen even now, and it is unlikely to occur in the future. The natural is frequently preferable to the artificial or the technologically assisted; not only this, the natural is generally much cheaper and is usually devoid of side effects. And of course, no technology is perfectly developed; there will always be associated problems, either inherent to the technology or to the way in which it is employed.

It is perhaps not surprising that those with scientific (and clinical) backgrounds on the one hand, and those with theological training on the other, tend to respond differently. This may reflect little more than contrasting disciplinary perspectives, and such differences are inevitable. But of course there is frequently overlap between the two since most of us do not live in hermetically sealed disciplinary compartments. Nevertheless, from time to time, tension between the two becomes evident, especially in relation to issues encountered at the peripheries of human existence, whether at the earliest stages or the latest stages. Far less debate surrounds the middle stages, even though these occupy most of our time as mortal beings, and most of the efforts of healthcare professionals. In addressing issues such as these I shall emphasize the importance of taking seriously the state of the science, neither ignoring it, nor embellishing it. As I wrote in *A Glass Darkly*, 'If theology is to engage meaningfully with scientific medicine it has to engage with medicine as it actually is, not with what it could become if it were to develop in a series of hypothetical, and possibly unlikely, directions'.[8]

7 Christian Institute, 'The Sanctity of Life', <http://christian.org.uk/briefingpapers/sanctityoflife.htm> accessed 29 May 2013; Philippa Taylor, *For What It Is Worth: The Status of the Human Embryo* (London: CARE/CBPP, 2002).
8 Jones, 'The Biomedical Technologies', 11.

In the same way, I take seriously theological input. I have no illusions about science; it requires direction. While Christian theology is but one source of such guidance, it is a source and one that Christians expect to be relevant in the contemporary world of science. And so I shall do my best to incorporate theological directives into my discussion, although I am well aware that I am not a theologian. Consequently, any theological proposals I bring to the table have an amateur feel to them. This is inevitable in a world where in academia the disciplines occupy their own well-defined territories and where there are discernible dangers in crossing from one territory to another. Interdisciplinary ventures are fraught ventures, and those who indulge in them can find themselves satisfying no one. However, no matter what dangers may be inherent in these ventures, it is vital that all who aim to speak relevantly in today's world cross these boundaries.

As a scientist I speak in the language of scientists and I think using scientific thought-forms. This may place me at a disadvantage since I am already biased towards acceptance of medical innovation and biomedical technological developments. My immediate reaction is to say 'yes' rather than 'no' in the face of all that is new. This may be correct, but of what relevance is this? If I had no scientific background, my inclination may be to say 'no' rather than 'yes'. Neither response carries any especial ethical or theological weight. We all bring with us certain disciplinary and cultural backgrounds, all of which have to be assessed on merit.

All need wisdom in determining how to live in a technologically dominated world. There is no escape from having to decide how a particular technology is to be used to help people rather than disadvantage them; when to use it rather than refrain from using it; what role it should play in the larger picture of healthcare within a society. Alternatively, one could reframe these considerations by adopting the premise that our starting point should be opposition to these developments on the ground that their very existence poses threats to humankind. In that case, there is no role for specific decision-making: no technology, no decisions required on how it will be utilized: end of discussion. In practice, however, this is far from the end of the discussion. No matter how much some groups oppose a particular technology in the biomedical area, it will be developed.

Refusal to acknowledge this is refusal to face up to reality, and to the serious decision-making that accompanies this.

From my Christian perspective, the challenge is to identify the character of the theological guidance that will assist decision-making. The context is provided by the science available, and the health needs of individuals and entire populations. In most instances, this guidance will not lead to definitive positions ('*the* Christian position'); rather, its aim is to adjust the tenor of the conversation by providing overall directions and goals. My role, as I see it, is to contribute to this conversation by pointing out the thrust of the technological imperatives, their strengths and their weaknesses, and the manner in which Christian stances can infuse it with rigour and hope.

Since I accept the legitimacy of much biomedical technology, I am committed to a search for directions for that technology and in so doing I look for theological guidance.[9] This leads to a quest to discover how theology might best contribute to a world dominated by biomedical intrusions. I take it as read that technology is developed in an attempt to help people rather than hurt them; that technology is useful under some circumstances but not others; and that dependence upon God can provide direction for decision-makers.

What then are we to expect of Christian input? In the title of this section I have deliberately used the terms a 'glass darkly' and 'uncertainty'. St Paul in writing to the Corinthian church was responding to the excessive enthusiasm of some of its adherents and the assured correctness of their views on a range of topics. In addressing these issues he reminded them that their current understanding is far from complete. Hence, the well-known reference to seeing through a 'glass darkly' or as in a 'mirror dimly.' 'For now we see through a glass, darkly; but then face to face: now I know in part; but then shall I know even as also I am known' (1 Cor. 13:12, *King James Version*). An alternative rendering is, 'Now we see only puzzling reflections in a mirror, but then we shall see face to face. My knowledge now is partial; then it will be whole, like God's knowledge of me' (*Good News Bible*).

9   Jones, 'The Biomedical Technologies', 12.

These words remind us that we too in the contemporary world are walking in uncertain territory. There is much that inevitably is ambiguous. It would be foolhardy to expect clarity on matters of such profundity, especially when we have been aware of the scientific details for such short periods of time. To make matters even more perplexing, the ground we are covering was unknown to any of the biblical writers, or even to any of the major theological figures whose guidance and judgement we are accustomed to seeking. It is my contention that it is incumbent upon us to steer a course between complete assurance as to the rightness of our cause and despair that helpful guidance will ever be found. I am only too aware that we do not see as clearly and unambiguously as we would like, and that all too often we are dealing with puzzling reflections rather than seeing with 20/20 vision. This applies to both the scientific and the theological aspects of the topics under discussion.

Theological insights

Since I am working within a Christian paradigm, the first place to turn is to Scripture. However, here one is immediately faced with a challenge of its own. It is deceptively easy to jump into biblical passages or even isolated biblical verses and emerge with definitive guidance on the specific bioethical issues that challenge us today. To avoid such an approach and provide a backdrop to this task, I shall listen initially to theological input.

The biblical scholar, John Rogerson, when seeking assistance to discover how far the biblical writers take us on the subject of abortion, writes:

> the use of the Bible is not a matter of selecting texts and of trying to apply them as though they were legislation for modern situations ... the Bible's primary function is to bring us to faith and to keep us in faith. The faith which we confess is faith in a God who responds to human need, who justifies the unrighteous and who seeks the outcast. The Bible lays upon us imperatives that derive from the heart of our

salvation, and our task is to work out those imperatives in the situation in which we find ourselves.[10]

In looking closely at abortion Rogerson freely admits that the Bible does not address in any direct manner the techniques available in modern societies. After all how could it when it was dealing with dramatically different cultural contexts? But he contends that it challenges us 'to include unborn children along with the defenceless and minorities whose task it is for the strong to defend.'[11] This immediately proves helpful, because while not directly addressing the question we wish to ask and have answered, it provides a broad framework within which we need to be doing our own detailed work as Christians. We will not find immediate direct answers, and this is a salutary starting point, but neither are we tempted to ignore the overall relevance of Scripture when deciding how to treat people.

A theological ethicist who has written extensively on medical ethics is Allen Verhey. Of his many contributions the one I shall initially refer to are his interpretive rules.[12] While the description 'rules' may be misleading, I think of them as 'directives', the suggestions are useful. According to Verhey, Scripture is to be read humbly, those using Scripture are to avoid interpretive arrogance, and they are to be aware that the context is that provided by the Christian community. Each one of these is pertinent to the task I have set myself.

The first directive calls for humility. Not one single individual or even scholar has all the answers on these bioethical quandaries, no matter to which Christian tradition they claim allegiance, no matter how definitive some of their forebears in the faith may have been, nor how categorical

---

10   John Rogerson, *Theory and Practice in Old Testament Ethics* (London: T&T Clark International, 2004), 98.

11   Rogerson, *Theory and Practice in Old Testament Ethics*, 98.

12   Allen Verhey, *Reading the Bible in the Strange World of Medicine* (Grand Rapids, MI: W. B. Eerdmans Publishing, 2003); Allen Verhey, *Remembering Jesus: Christian Community, Scripture, and the Moral Life* (Grand Rapids, MI: Wm. B. Eerdmans Publishing, 2002).

their church hierarchies may be today. Not one of these has assured answers on every minute point raised by current bioethical debate.

At first this appears self-evident, and yet in practice it is less so. The reason is that many Christian commentators come out of traditions with well-defined dogmatic foundations, the truth of which is of immense importance to the validity of that tradition. These are first-order statements about the nature of God, the person and work of Christ, the nature of the church and the Kingdom of God both now and in the future. No matter what debate there may be on detailed interpretation of some of these core doctrines, there is also an assurance about them on the basis of specific Scriptural teaching. By contrast, the character of second order issues, such as those encountered in contemporary bioethical discussion, is far more uncertain with far fewer points of definitive agreement.

This leads into the second of Verhey's rules or directives, namely that we are to avoid interpretive arrogance. One too often comes across assertions that claim, for example, that since God saw and knew the psalmist prior to his birth (as in Psalm 139), the human 'soul' must be present from conception. Or again, the stance that all human beings are made in the image of God must apply from conception, and therefore all embryos image God and have the right to life.[13] There are no doubt valuable theological insights in these statements, and ones that Christians can readily affirm, but intermixed with these are questionable bioethical assertions that go far beyond any views that the biblical writers could possibly have had in mind. Detailed assertions along these lines owe far more to non-Scriptural (in some cases, to traditional) viewpoints than they do to guidance emerging from Scripture itself.

The third guidepost provided by Verhey is that there is no room for the privatization of ethical deliberation. He argues that biblical scholars, theologians and ethicists are not by themselves competent to determine a Christian response without reference to a (the) wider Christian community. I take it that the wider community will include the input of scientists,

---

13   B. Waters and R. Cole-Turner, eds, *God and the Embryo: Religious Voices on Stem Cells and Cloning* (Washington, DC: Georgetown University Press, 2003).

clinicians, lawyers, economists and policy makers, let alone non-expert people of faith. We need each other within the community of faith; scientists and clinicians need theological scholars, but the latter also need scientists, clinicians and many others.

One outcome of this third point is that any statements and ideas on bioethical matters from within the Christian community should be consonant with accurate and up-to-date scientific concepts. The intention of this is not to give primacy to the science – that would be unwarranted – but to ensure that any theological discussion is informed by relevant scientific ideas rather than by outdated ones. This in turn highlights the moveable nature of science-theology dialogue in these areas; from this there is no escape. It also highlights the danger of theological viewpoints that are allegedly based solely on Scriptural notions, but owe far more to scientific ideas than anyone suspects, and unfortunately often to outdated ones.

Verhey has expanded considerably on these matters elsewhere in his writings, where he delves into what he describes as the strange world of sickness in Scripture.[14] He writes, 'In memory of Jesus and in hope the Christian community will delight in human flourishing, including the human flourishing we call "health", but also be able to endure even the diminishing of human strength we call "sickness" with confidence in God'.[15]

For Verhey, our remembering Jesus and his attitudes will dispose us towards a number of crucial attitudes of our own: respect for the embodied integrity of people, for their freedom and identity, the need to nurture community, and to support and care for – and if feasible – cure the sick.[16] On the other hand, he also stresses the limited nature of our powers that are far from being messianic. Hence we are not to have extravagant expectations of any human power, including medical powers, and they are certainly not

---

14   Allen Verhey, 'What Makes Christian Bioethics Christian? Bible, Story, and Communal Discernment', *Christian Bioethics* 11/3 (2005), 297–315.
15   Verhey, 'What Makes Christian Bioethics Christian?', 305.
16   Verhey, 'What Makes Christian Bioethics Christian?', 308.

to be idolatrous.[17] This latter point is of especial importance for biomedical technology that routinely exaggerates the reach of the scientific enterprise and its ability to extend the capabilities of humans.

The counterbalance advocated by Verhey emanates from the 'not yet' character of our life and medicine. Moral ambiguity is inescapable, since we will never be able to bypass having to cope with good ends that come into conflict with one another at the same time as we have to confront evil ends. As he writes, 'The memory of Jesus does not provide any neat and easy resolution to such conflict. It does not usher in a new heaven and a new earth, either. Here and now there is ambiguity'.[18]

While the balance enshrined here is readily accepted within Christian thinking, it tends to meet a hostile reception in some other circles. As we shall encounter when dealing with transhumanism (chapter 8) the concept of inherent limits to human endeavours and abilities is anathema.[19] It is at this juncture that the modern medical enterprise encounters the clash of worldviews, and hence dissonance for Christian thinkers.

The emphasis of Christian writers on the limited abilities of humans and their non-messianic character comes to the fore time and time again. Andrew Goddard has argued that techniques should not become 'idols around which we create an alternative salvation-history and drama of redemption to that revealed in Scripture'.[20] For him biomedical technologists should never take on the aura of saviour, since they lack any such powers.

In similar vein Ian Barns argues that techno-utopians are wrong in the belief that human mortality can be transcended through technology.[21]

---

17   Verhey, 'What Makes Christian Bioethics Christian?', 311–12.
18   Verhey, 'What Makes Christian Bioethics Christian?', 313.
19   Nick Bostrom, 'Transhumanist Values', *Review of Contemporary Philosophy* 4/1–2 (2005), 87–101.
20   Andrew Goddard, 'The Place of the Bible in Medical Ethics', in D. Gareth Jones and R. John Elford, eds, *A Glass Darkly: Medicine and Theology in Further Dialogue* (Bern: Peter Lang, 2010), 133–56, 154.
21   Ian Barns, 'Debating the Theological Implications of New Technologies', *Theology and Science* 3/2 (2005), 179–96.

His concern is that the association between technology and transcendence animates major areas of technoscience, including genetic engineering. He argues that 'the project of transcendence through technology is unsustainable and a destructive folly, and that human life needs to be lived gladly within the limits and diverse possibilities of our existing material condition.'[22]

His Christian response is the need to image Jesus in the sense that we do not aim to project human power and control, but faithfully trust God and become a suffering servant. While he sees the task of critiquing excessive reliance upon technological powers as crucial, the central task from his angle is not just to recognize the limits of nature but to seriously address what it means to be human in a posthuman(ist) culture.[23]

For him this is relevant in the biomedical area, since the influence of Jesus was felt in his acts of healing, exorcism and control over nature, and therefore the model he espoused serves as the source of freedom from the crippling effects of disease, ignorance and spiritual darkness.[24] In other words, even though we have to be cautious in the face of technological excesses, we also have to take seriously the contribution of science and technology to Christian imperatives.

These directives are general in nature, but how do they help in the murky world of contemporary biomedicine? It would be unfair to expect them to provide precise answers to precise ethical queries at the interface between science and ethics; that is not their intention. What they do is provide guidance, and it is this guidance that people such as myself need to interpret when faced by focused dilemmas.

Neil Messer has provided a step is this direction with what he describes as four diagnostic questions:[25] He outlines them in the following manner.

22   Barns, 'Theological Implications of New Technologies', 185.
23   Barns, 'Theological Implications of New Technologies', 185.
24   Barns, 'Theological Implications of New Technologies', 191.
25   Neil Messer, 'Christian Engagement with Public Bioethics in Britain: The Case of Human Admixed Embryos', *Christian Bioethics* 15/1 (2009), 31–53.

- Is the project good news to the poor, the powerless, those who are oppressed or marginalized in any way?
- Is the project a way of acting that conforms to the *imago dei*, or is it an attempt to be 'like God'?
- What attitude does the project manifest towards the material world (including our own bodies)?
- What attitude does the project manifest towards past human failures?[26]

In his book *Respecting Life* Messer adds a fifth diagnostic question: 'What attitude does the project embody towards our neighbours?'.[27] For him, this is a central theme in Christian ethics, and he seeks to apply love of neighbour to a wide variety of groups, including embryos.

One example of how Messer takes these questions further is provided by the following quotation with respect to the third one. 'A more theologically adequate attitude to the physical will ... recognize the material world as good and worth taking trouble over, yet flawed and in need of transformation and will not make the mistake of thinking that physical life in the present material world exhausts what humanity has to hope for'.[28]

Frustrating as these questions might be for those intent on immediate answers, they complement the previous considerations. All of them in their various ways are attempting to work within the context provided by the biblical material, and especially by the ministry of Jesus. Just as importantly, they recognize the role members of the Christian community in the twenty first century have to play in interpreting them as they address issues facing them in laboratories, hospital wards, fertility clinics, and the spheres of governmental policy making.

These theological pointers provide salutary reminders to all – regardless of worldview – that the technologies with which we are concerned are to benefit real people including those in the majority world. Justice,

26   Messer, 'Christian Engagement with Public Bioethics', 41–3.
27   Neil Messer, *Respecting Life: Theology and Bioethics* (London: SCM Press, 2011), 40.
28   Messer, 'Christian Engagement with Public Bioethics', 43.

equality and dignity demand nothing less. They also call us to look back and recognize where there have been failures, where the bright hopes of technological triumph have come to naught, and where they have been abused to the detriment of those they were supposed to help.

We are also reminded to look beneath the surface of self-assured pronouncements and assess whether they adhere to guidelines such as these. It is this adherence that is the lynchpin of a Christian response, rather than agreement over specific positions, which for all their clarity may or may not represent a faithful outworking of theological principles like these.

The theological commentators I have referred to take the writers of Scripture seriously and seek to interpret what they were stating into the thought forms and challenges of the contemporary world. They do not seek to find verses or statements in the biblical text that they then apply directly to a contemporary bioethical quandary. Examples of the latter approach crop up repeatedly in discussions on abortion, the commencement of human life at conception, and homosexuality, and by implication provide clear directions for thinking about the moral status of the embryo, contraception, the sanctity of human life and euthanasia.

Apart from the obvious comment that the writers encountered in Scripture were writing 2,000 years ago, and in some cases far more than that, they cannot be expected to provide specific guidance on issues that would have been totally foreign to them. In the same way commentators in the early years of the nineteenth century were equally ignorant of genetic and neuroscientific issues. They lacked the language and the basic conceptual schemes to contribute. Certain general principles may of course prove useful today, but that is exactly what the theological commentators earlier in this chapter are seeking to provide. What then of the Bible? What can we glean from its writers?

I consider one can divide possible approaches into a number of categories:[29]

29   D. Gareth Jones, 'Responses to the Human Embryo and Embryonic Stem Cells: Scientific and Theological Assessments', *Science & Christian Belief* 17/2 (2005), 199–221.

1) the Bible alone provides a complete guide to ways in which Christian decision-making should be framed, making scientific input irrelevant;

2) the Bible is one of a number of sources of concepts and information, but is the major determinant whenever there is conflict or confusion;

3) the Bible is one of a number of sources of concepts and information, and helps to inform decision-making, but may not be the major source;

4) the Bible is irrelevant and hence can provide nothing of any interest to scientists or ethicists.

Christian writers are found within each of the first three categories. Category 4 reflects the stance of those whose starting point lies outside the Hebraic-Christian tradition, and hence is irrelevant for my purposes. In order to take this discussion further I shall introduce a real life scenario, that grounds much of the more theoretical debate of the previous pages. I shall return to the three categories open to Christians in the discussion of the following scenario.

Face to face with medical decision-making

Imagine the following real-life scenario. A couple had two new grandchildren, born one day apart, a girl to one of their children and then a boy to another of their children. The girl was everything one would expect of a lovely newborn child – the joy at the birth of a 'perfect' new human being. She was self-evidently a gift of God; delightful and everything one could wish for. And she has gone on developing in this way, bringing immense joy to her parents.

The birth the following day of a little boy was initially accompanied with similar feelings, and yet these were soon overtaken when he had to have a major operation within two days of birth. This immediately suggested the diagnosis of cystic fibrosis (CF), a diagnosis confirmed by genetic analysis a couple of weeks later. The treatment he received at the time and has subsequently received has been state-of-the-art, and his progress has been as good as could be expected. But how should the grandparents respond as Christians?[30]

The contrast between the two births was dramatic, as was the ongoing trajectory of natural growth and development in the one case, over against intense therapy-dependent growth and development in the other. Healthy development in the one is what one would usually consider to be 'normal'; in the other it is 'artificial' or perhaps more accurately 'artificially supported'. And yet the ideal for each of the children is to live as productive and fulfilled lives as possible, even though there will always be far more limitations in the one case than in the other.

Their grandson with CF is intimately dependent for his daily needs upon a very high level of technological expertise.[31] If the facilities had not been available, he would have died within a week of birth. The resources necessary to save the life of a child like this, and hopefully provide him with an ongoing good quality of life, are expensive and essential. There can be no escape from sophisticated science and technology in this case; in the absence of this assistance there will be no ongoing life since illness and premature death will intervene. In this context science and technology are contributing to the enrichment of humanity.

Research is central to the treatment of CF and therefore to the wellbeing of those with CF. The life expectancy of those with this genetic condition has been transformed over the past twenty to thirty years. Inevitably, this has profound implications for patients and family alike, since they impact directly on the life hopes and aspirations of a real human being. It is also salutary to reflect that had this been one of the couple's children, rather than one of their grandchildren, the outlook would have been very

---

30   Jones, 'Conclusion', 234–7.
31   P. B. Davis, 'Cystic Fibrosis since 1938', *American Journal of Respiratory and Critical Care Medicine* 173/5 (2006), 475–82.

much bleaker (just as it is today in far too many countries). And this, of course, provides only a tiny glimpse into a multi-dimensional issue – that the lives of many people are hugely dependent on the sophistication and availability of medical treatment.[32]

The parents are utterly devoted to this son, and commit themselves to his welfare each day, no matter how many hospital visits are necessitated nor how much it interferes with their daily schedules. Their love for him is unbounded. But what of the future? As the parents look ahead, they begin to contemplate the birth of a second child. They would prefer not to have another child with this condition because the demands on them, let alone on the affected child, are immense. They have *four options*.

Option 1. They could decide against having another child; the one-in-four prospects of another child with CF are more than they can bear. Neither do they want to bring into the world another individual with a debilitating condition like CF.

Option 2. They could take a chance and proceed with a pregnancy as normal. They know that their next child has a one-in-four chance of having CF, but they hope it will be unaffected. Whatever eventuates they will continue with the pregnancy even if it means having a second child with CF. If they are Christians they might argue that the welfare of their next child is in God's hands regardless of the outcome.

Option 3. They could take a chance, knowing that they can have an abortion if the fetus when tested at nine to ten weeks' gestation using chorionic villus biopsy turns out to be affected. They would only go down this path as a very last resort. They might feel that to abort a fetus with CF would be to demean what they value about their little boy.

Option 4. They could go in the direction of IVF (*in vitro* fertilization) and PGD (pre-implantation genetic diagnosis); see chapter 3. If this dual procedure is carried out and shows that an embryo at the four- or eight-cell stage does not have the CF gene it will be transferred to the woman in the normal way. On the other hand, if the CF gene is present in an embryo that embryo will be discarded and the same procedure will be carried out

32   Jones, *Designers of the Future.*

on a second embryo and so on until an embryo lacking this gene is found. The couple is prepared to consider this as a possible way forward but feel they are treading a very delicate and uncertain path spiritually.

For this couple there is no escape from the reality of CF, and its profound effects on any children they bring into existence. They know they do not have to take the technological option, and no one is pressuring them to do so. The one thing that is assured is that they have to make a choice, and having done so will have to live with the repercussions and consequences of their choice. Depending on the direction they take, they could decide to have no further children, they may end up with a child with CF, or a child without CF, and they may generate embryos or fetuses that will be discarded as embryos or aborted as fetuses. These are invidious choices, but there is no escape for a couple in this situation.

When faced with such a personal and taxing dilemma as this, one has to sympathize with those who feel that it taxes their Christian convictions to almost breaking point. The competing forces at work within the dilemma pit one life against another, one set of values against another set. And where does the Bible enter the picture? One answer is to aim to protect the embryo or fetus at all costs, and this is frequently the answer given by Christians. But again, is this perspective nearly as firmly based upon unequivocal biblical teaching as commonly assumed by some? To explore this further we should return to the three categories touched on in the previous section.

Bible input and theological perspectives

According to *Category 1* the Bible alone provides a complete guide to the ways in which Christian decision-making should be framed, making scientific input irrelevant. In this instance, the diagnosis, prognosis and treatment are all bound up in an understanding of a particular genetic condition. There is no getting away from scientific input that demands responses. It is very difficult to see how the Bible taken in isolation of scientific input can possibly provide the only relevant input for decision-making in these circumstances. Perhaps more to the point, what is the character of any biblical input?

Categories 2 and *3* are far more plausible, with their stress on the Bible as one of a number of sources of concepts and information. The driving force then becomes the balance between biblical and scientific inputs, and the extent of any biblical input. Whichever emphasis one wishes to espouse, the reality is that any treatment will be scientific in nature. The way in which this science is used will stem from the attitudes and aspirations of the parents and this is where the Christian character of the decision-making will come to the fore. What then of the four options open to this young family?

Option 1 raises issues to be encountered in chapter 2, that of childlessness, only in this case it is self-inflicted – albeit for extremely wellintentioned reasons. Is this an acknowledgement that God is in charge, or are these human beings taking too much control into their own hands? Whatever answers are given will have considerable repercussions for one's views on contraception.

Option 2 with its one-in-four chance of having a child with CF is an acceptable way forward for Christians with its explicit recognition that God can use both illness and ongoing limitations for good. It has many merits, but is it automatically the only or even the ideal way for Christians? By rejecting some of the possibilities opened up by scientific advance, it is prepared to allow into the world an individual with a well-recognized disease. A cautious stance like this one cannot be held up as an exemplar of biblical teaching since the link between it and the attitudes of Jesus is exceedingly unclear. While the rights and wrongs of such ways of acting are hotly contested, there is no explicit Scriptural warrant for undertaking them. More positively, it may result in the birth of a child without CF, much to the delight of the parents, but even if this is the welcome outcome, it has been taken in the face of a major risk that could theoretically have been avoided.

Option 3 will be the most problematic for most Christians with its willingness to abort a fetus with CF. It may also have severe negative repercussions for the parents, and especially the mother. It also gives the appearance of being life-denying. This option revolves around the ethical and theological legitimacy or otherwise of induced abortion. This is not the place to discuss the merits of abortion for genetic defects, except to state that in my view there is not clear biblical warrant one way or the other.[33] The interests of each member of this family, including the potential or future interests of the fetus, have to be assessed in light of the picture presented by the New Testament writers of the body of Christ.

What then of *Option 4*, with its dependence upon IVF and PGD? Quite apart from the financial costs involved, it is invasive and involves the selection of embryos. Many Christians reject this possibility with its reliance upon choosing one embryo over another, and inevitably destroying embryos.[34] Destruction of embryos is unacceptable to many Christian writers, with one referring to this as 'hatred of the material creation'.[35] The debate so often comes down to the value placed upon embryos to the neglect of the other participants including the future child. Integral to the scientific milieu in which we live are techniques like IVF and PGD, and their availability has to be taken into consideration in deciding how best God is glorified in specific situations like the one here.

As the parents contemplate the options, they will need to take account of injunctions to protect the defenceless and disenfranchised, the importance of human flourishing and their ultimate dependence upon God. They will be aware that Jesus came to proclaim good news to the oppressed, and that human life is not devoid of meaning simply because it is physically flawed.

Alongside this they will be reminded of the transformative power of the physical and spiritual healing that Jesus brought and that can be experienced today in the midst of Christian community. They will also be grateful for the manner in which human creativity manifested in medical

33 D. Gareth Jones, 'Abortion: An Alternative to the Conflict Paradigm', in B. Richards and V. Pfitzner, eds, *Issues at the Borders of Life* (Adelaide: ATF Press, 2010), 11–21; D. Gareth Jones, *Brave New People: Ethical Issues at the Commencement of Life* (Leicester: Inter-Varsity Press, 1984); Jones, *Valuing People*.

34 M. Junker-Kenny, 'Genetic Perfection, or Fulfilment of Creation in Christ?', in C. Deane-Drummond and P. Scott, eds, *Future Perfect? God, Medicine and Human Identity* (London: T&T Clark, 2006), 155–67.

35 Messer, *Respecting Life*, 112.

achievements brings hope and new prospects in the midst of illness. However, uncertainty and ambiguity are never far from the surface, and the limited nature of human powers should never be downplayed.

Considerations like these do not lead inevitably in any one direction, but they do constitute the environment within which decision-making should occur. They need to be worked out in terms of what is in the best interests of the family as a whole, the existing child with CF, any future children who may exist with or without CF, and possibly any embryos. The weight placed on each of these will depend on many factors, and the balance achieved between these often conflicting interests should reflect the diverse relationships that characterize the family, and human and church communities.

Any decision the parents take will be an agonizing one; this applies in varying degrees for all four options. If they go down the IVF and PGD route, they will be agreeing to dispose of the affected embryos. The easy way out of this dilemma is to remain ignorant; they hope to avoid having to face these options. Unfortunately, this is not completely possible even if they decide against the IVF/PGD direction, since they are still confronted by the choice of refraining from having another child and that of going ahead and taking a risk. They are to some extent shielded from making a difficult, and possibly invidious, decision; if they want another child they will take what comes. That may of necessity have been the case thirty years ago, but it is no longer so. Technology makes certain things possible, as does genetic knowledge, but with this come inevitable ethical choices.

Is there a biblical mandate for remaining ignorant? This takes us beyond the narrow biomedical queries into much more general territory. Is freely chosen ignorance a Christian virtue, or is it a denial of the biblical mandate to be creative, to fill the earth and replenish it, and to be stewards of God's good creation?

I remain to be convinced that ignorance in and of itself is an acceptable way forward for Christians. If ignorance is not a moral virtue it follows that option 2 is not as virtuous as frequently depicted. Few would consider ignorance to be a virtue when confronted by malaria, tuberculosis or dysentery, let alone by measles or smallpox, about which something can be done in all these cases. Neither would they consider these to be part of

God's good creation. Failure to act in these circumstances, a far too frequent occurrence in many of the countries of the contemporary world, is an evil that blights the world of which we are part. Genetic advance is increasingly forcing us to confront and then make hard choices, and the parents in this scenario find themselves in precisely this predicament.

In considering the four options I have oscillated between them as I have looked at the pros and cons of each. But are there any additional biblical directives that might be of assistance? For instance, how did Paul deal with some of the applied ethical issues that faced the early church? Eating food sacrificed to idols is one example (1 Cor. 8). What were these early Christians to do? There were arguments on both sides, in favour and against, and different groups responded in one or other way.

The issues of course were hardly the same as whether or not to use a technological procedure like IVF/PGD to circumvent a genetic condition, and yet there is a common underlying theme. In addressing his issue, Paul drew a distinction between knowledge and love: 'knowledge puffs up, while love builds up'. Their knowledge of course was incomplete, and they could all too readily become too assured that they had the correct answer. They had to exercise care and discernment, within a context provided by love and concern for others – especially for those who adopted the opposite perspective.

With this in mind we should not rely on a technological solution to the exclusion of other considerations. Technological knowledge puffs up and may make us too dependent upon our own wisdom and our projects. While such ways, including the use of IVF/PGD, are not out of bounds for the Christian believer, they should be resorted to only when we have assured ourselves that at the most fundamental level we are dependent upon God, and trust him no matter what eventuates. Love and concern for others are essential ingredients within all ethical decision-making.

What this means in practice is that we do not resort to a technological solution without very considerable reflection and without putting its use into a broad human and divine framework. On the other hand, ignorance by itself is not a virtue. It always has to be accompanied by trust in the goodness of God, and committing oneself to the purposes of God who has one's welfare and the welfare of humanity at heart. A technological direction as an end-in-itself will end in dehumanization; ignorance as an end-in-itself leads to fatalism. In Christian terms, a solution isolated from trust in God and his good purposes will prove misleading, and will lead away from God.

Science and faith are inextricably linked. Not one of the various options put forward is untenable, as long as each is adopted within a framework governed by prayer and a realistic assessment of the technology involved and its alternatives. However, it is unlikely that all options will be regarded as ethically equivalent by different groups of Christians. The underlying context is provided by the knowledge that elimination of CF is not a viable project at present, although a disease like smallpox has been essentially eliminated. A far more realistic priority is continued research to improve the treatment of those with CF. The far-reaching question is whether elimination of the CF gene would involve excessive shaping of our bodies and identities or represent a triumph of medical and human creativity. The manner in which we answer this will depend upon our interpretation of the mandate we have for reconstructing and controlling the human body as stewards in God's world.[36]

Concluding comments

My basic presupposition is that there is guidance to be found in the Bible, guidance that will assist those who wish to act as Christ's followers in the contentious and highly problematic world of modern medicine. By its nature any biblical guidance is at a general level, which is not to demean its value. However, it is not always what Christians are looking for, since it leaves a great deal to the judgement and discernment of individuals and

---

36   Ted Peters, *Playing God? Genetic Determinism and Human Freedom* (New York: Routledge, 2003); B. R. Reichenbach and V. E. Anderson, *On Behalf of God: A Christian Ethic for Biology* (Grand Rapids, MI: Eerdmans, 1995).

communities. That is probably as it should be, but are there in addition more precise and finely honed directives in Scripture, as one finds in the writings of Christian and other pressure groups, and also in the public pronouncements of Church denominations and the Vatican?[37] This is the world of rules and regulations, with their specific do's and don'ts, and often framed within a context of cautious prohibitions.

I have steered clear of any approach based on searching for simple formulae, since these stem far more from extra-biblical thinking than from anything inherent within Scripture. If we are not very careful we can end up unintentionally misusing the Bible, by expecting of it answers its writers never claimed to provide. By taking as my central example a genetic condition that intrudes massively into the lives of all concerned by making considerable demands on a range of related parties, we have encountered the lack of clarity so often encountered in biomedical illustrations, with their interplay of human and scientific factors. How one acts as a Christian in the midst of these tensions and pressures depends upon biblical interpretation and theological reflection. Christians will seek to listen to and follow Jesus, no matter how foreign the contemporary world is to his. Inevitably, different Christians will make different decisions in such contentious and unclear territory; that is a reality of bioethical and theological debate. But there is no reason why Christians and likeminded people should not respond positively and even enthusiastically to the challenges of modern medicine, as long as these are always viewed alongside the central importance of the character and integrity of the participants.

37   See, for example, Congregation for the Doctrine of the Faith, '*Donum Vitae*, Instruction on Respect for Human Life in Its Origin and on the Dignity of Procreation: Replies to Certain Questions of the Day', (1987) <http://www.vatican.va/roman_curia/congregations/cfaith/documents/rc_con_cfaith_doc_19870222_respect-for-human-life_en.html> accessed 28 March, 2013; Congregation for the Doctrine of the Faith, 'Instruction *Dignitas Personae* on Certain Bioethical Questions', (2008) <http://www.vatican.va/roman_curia/congregations/cfaith/documents/rc_con_cfaith_doc_20081208_dignitas-personae_en.html> accessed 4 April 2013; Southern Baptist Convention, 'Resolution on Genetic Technology and Cloning', (June 1997) <http://www.sbc.net/resolutions/amResolution.asp?ID=571> accessed 29 May 2013.

A story of two domains

Contrasting domains

As a broad generalization Christian writers tend to be more concerned about the reproductive realm than the neuroscientific one. There may be a simple explanation: it is familiar territory and is relatively easy to understand what is involved when embryos are manipulated or artificial processes replace normal reproductive ones. We are dealing with the nature of human life and what makes us what we are. By contrast intrusions into the brain and into mental processes take us into very mysterious and complex intellectual territory, the ethical dimensions of which may be far less easy to discern. For many the brain is little more than a black box – indecipherable and unknowable. The question confronting us is whether either or both represent illicit forays into a realm that may lie outside the bounds of legitimate human activity.

I shall argue that, even though the dimensions of the two areas give the appearance of being very different, the tensions and complexities are remarkably similar. They may even be greater and more immediate in neuroscience, since numerous means of modifying people's brains are already with us – they constitute present reality. This is not to deny that at the beginning of life momentous choices may, on occasion, be made between one nascent human life and another. By contrast the neuroscientific realm is more concerned with altering the functioning of an existing human being, or perhaps seeking to extend it in some way. While its dimensions appear to be more limited, numerous individuals may be affected with consequences for all their lives.

In order to dissect the theological ramifications of both areas I shall look at two contrasting illustrations in each case. Intrusions in both spheres represent the exertion by human beings of significant influence over human life. From a Christian perspective both represent potential means for glorifying God and serving humanity, or alternatively for asserting mastery over human existence. Each area will be taken up in greater detail in subsequent chapters.

Reproductive biology:
Using technology to bring new life into existence

Discussions at the beginning of human life almost invariably centre on *in vitro* fertilization (IVF) and its myriad off shoots. Important as these are (see chapter 3), I shall set the scene by looking at other technological means of helping those with fertility problems, since these in their turn introduce relevant considerations.

Condition 1 – polycystic ovary syndrome

> Janine is twenty-six years of age and has been married for three years. She and her husband would like a child but are having difficulties conceiving. After consulting a gynaecologist it soon emerges that Janine is suffering from polycystic ovary syndrome (PCOS).

This relatively common condition affecting 5–10 per cent of women of reproductive age, occurs when the woman's eggs do not mature in their follicles and are not released from the ovaries.[1] Instead, they form very small cysts in the ovary. The condition is related to an imbalance of a woman's

---

1   R. J. Norman et al., 'Polycystic Ovary Syndrome', *Lancet* 370/9588 (2007), 685–97.

female sex hormones (oestrogen and progesterone), and while the causes are unknown, insulin resistance, diabetes and obesity are all strongly correlated with the condition. It may also have a genetic component.

Polycystic ovaries develop when the ovaries produce excessive amounts of male hormones (androgens), particularly testosterone. These elevated insulin levels contribute to, or cause, the abnormalities seen in the hypothalamic-pituitary-ovarian axis that lead to PCOS.[2] Treatment includes losing weight, the use of progesterone pills to help make menstrual cycles more regular and taking Clomiphene (Clomid) or FSH (follicle stimulating hormone) to help overcome infertility. Other possibilities include Metformin and other insulin-sensitizing drugs in an effort to decrease the blood level of insulin that in turn may help to counteract the underlying cause of PCOS.[3] This is relatively straightforward medical treatment, the goal being to overcome an obstacle that is hindering fertility. The array of treatments available is directed at rectifying hormonal imbalance, since it is this that appears to be causing the problem.

What we are dealing with in this illustration are technological means of alleviating childlessness. There is no way we can look to the biblical writers for comment on procedures such as these, although childlessness appears on a number of occasions in Scripture. Importantly, it is never presented as positive, or even acceptable: it is always a tragedy. In 2 Samuel 6:23 the death of Michal, Saul's daughter, is accompanied by the narrator's comment 'she had no child until the day of her death'. The expectation was that women were to bear children, and failure to do so was a reason for mourning.

This is hardly surprising since the blessings of fertility appear powerfully throughout Scripture. From the very beginning there is the command to be fruitful and multiply and fill or replenish the earth (Gen. 1:28; 9:1), while the psalmist refers to God settling the childless woman in her home as a happy mother of children (Ps. 113:9). Repeatedly, there are examples

---

2   Adam H. Balen and Anthony J. Rutherford, 'Managing Anovulatory Infertility and Polycystic Ovary Syndrome', *BMJ* 335/7621 (2007), 608–11.

3   Thomas Tang and Adam H. Balen, 'Use of Metformin for Women with Polycystic Ovary Syndrome', *Human Reproduction Update* 19/1 (2013), 1.

of God causing previously 'barren' women to become pregnant. Consider Abraham and Sarah (Gen. 16–18, 21), Isaac and Rebekah (Gen. 25:21), Jacob and Rachel (Gen. 30, 35), Manoah and his wife (Judg. 13:3–5), Elkanah and Hannah (1 Sam. 1), the Shunammite woman (2 Kgs 4:8–17), and Zechariah and Elizabeth (Luke 1). God is recognized as the one who opened and closed the womb (Gen. 16:2; 20:18; 1 Sam. 1:5) while conception after a long period of infertility was a cause for rejoicing since God had removed the woman's reproach (Gen. 30:23; 1 Sam. 1:10–11; 2:1–10).

Nevertheless, some of those in the Old Testament went to considerable lengths to circumvent childlessness.[4] While these are strikingly different from anything encountered today, with their use of maidservants (Sarai uses Hagar Gen. 16:2, Rachel uses Bilhah, and Leah Zilpah Gen. 30:3, 9) and levirite marriage (Deut. 25:5–10), they were examples of searching for remedies to overcome infertility. The goal in all instances appeared to be to strengthen the family unit, even if some of the attempts appear to have been misguided. This negative image of childlessness is communicated by promises which declare that if Israel is faithful to God, there will be no barrenness in the land (Exod. 23:26; Deut. 7:14). On the other hand, childlessness was sometimes seen as the result of sin and/or divine displeasure (Gen. 20:18; Num. 5:11–31; 2 Sam. 6:20–23).

In modern writings Christian commentators have dwelt mainly on the validity or otherwise of artificial means of contraception, and on issues raised by embryo-based artificial forms of fertilization such as IVF and its many offshoots. Only limited attention has been paid to the validity of other procedures to bring about fertilization. Perhaps this reflects a lack of theological concern with these procedures, or an unwillingness to confront them. The Scriptural evidence points in a positive direction, and yet great care has to be exercised in moving directly from these specific examples of God's dealings with his people within a very particular religious setting, to addressing a range of technological solutions across whole populations in pluralist cultures.

---

4   D. Gareth Jones, *Manufacturing Humans: The Challenge of the New Reproductive Technologies* (Leicester, England: Inter-Varsity Press, 1987).

In general terms I support therapeutic efforts at alleviating infertility such as that in the illustration of PCOS, since the goal is to bring wholeness to marriage, and the husband and wife relationship. For Christians the framework is that of dependence upon God both individually and communally, recognizing the importance of the support and understanding of other members of the Christian community, and of Christianity's affirmation of the value of the marital bond, with or without children. This does not preclude undertaking serious efforts at overcoming childlessness, although good practice would suggest that minimally invasive interventions should precede more aggressive ones.

A Christian perspective accepts that there are limits to human overcoming, since the hope of the gospel is an eschatological hope that looks to the fulfillment of God's purposes in the renewal of all things – our bodies and relationships included.[5] For a couple embroiled in this situation it will be important to arrive at the position that, no matter how important it is to have one's own children, this is less important in the long run than obedience to the call of God. They are to learn that a technological solution is to be adopted within the context of their ultimate trust in, and dependence upon, the goodness of a loving God and Saviour. Consequently, they, and the community around them, are to seek God's help and directives in all the challenges brought about by infertility and possible childlessness.

Condition 2 – severe fetal abnormality

Jess and Steve are expecting their first child and are filled with hopeful anticipation at all that this new life will bring to them as a couple. They are already thinking of names for the child and in their imagination are beginning to map out this new individual's life course. It is in this spirit that Jess has a routine scan, since everything with

---

5   Ted Peters, 'Resurrection of the Very Embodied Soul?', in Robert John Russell, et al., eds, *Neuroscience and the Person: Scientific Perspectives on Divine Action* (Indiana: University of Notre Dame Press, 1999), 305–26; N. T. Wright, *Surprised by Hope: Rethinking Heaven, the Resurrection, and the Mission of the Church* (New York, NY: HarperOne, 2008).

the pregnancy has so far been routine and uneventful. However, instead of coming away from the appointment with reassuring news, she is informed that something is wrong – the fetus' heart is not developing properly; and the situation looks serious. While there is little doubt that development is abnormal, the severity of the situation needs to be investigated further, necessitating an appointment at a specialist centre in another city. A few weeks later she has this appointment, and the detailed scan reveals major abnormalities with the developing heart, and the couple's worst fears are confirmed. The prognosis is that if nothing is done the child will probably live for a few days after birth; if major repeated surgery is undertaken the child may be able to live for up to four or five years and will have an uncertain quality of life.

This scenario differs from the first in that there are no fertility problems; the couple has had no problems in conceiving. Unfortunately, the development of the fetus is far from normal, and the prospects of a reasonably normal life for the new individual are minuscule. The organ at fault in this case is the heart, although any of the major organs could have been involved, depending on when and how embryonic development went awry. The causes do not concern us, nor do the details of the heart defect. Major pathologies like this are relatively common, although most of them are 'aborted' spontaneously very early in pregnancy, a phenomenon referred to as pregnancy wastage. Estimates put the frequency of pregnancy wastage at around 60–70 per cent, with 99 per cent of this wastage occurring during the first trimester of gestation.[6] The major contributing factor to this astronomically high rate of pregnancy wastage is that of chromosomal abnormalities. The frequency of these in spontaneous abortions has been reported to vary from 8 to 83 per cent,[7] the most frequently quoted overall figures being 40 to 50 per cent.[8]

6   N. S. Macklon, J. P. Geraedts, and B. C. Fauser, 'Conception to Ongoing Pregnancy: The "Black Box" of Early Pregnancy Loss', *Human Reproduction Update* 8/4 (2002), 333–43.

7   C. M. Strom et al., 'Analyses of 95 First-Trimester Spontaneous Abortions by Chorionic Villus Sampling and Karyotype', *Journal of Assisted Reproduction and Genetics* 9 (1992), 458–61.

8   G. C. Wolf and E. O. Horger, 'Indications for Examination of Spontaneous Abortion Specimens: A Reassessment', *American Journal of Obstetrics and Gynecology* 173 (1995), 1364–8.

The relevance of these figures for the couple in this scenario is that there are natural mechanisms for eliminating the majority of chromosomal anomalies during pregnancy, and these are extremely efficient. For instance, it appears that over 99 per cent of chromosomal abnormalities are eliminated through spontaneous abortion or fetal death.[9] While there are different categories of chromosomal abnormalities, the overall message is that very few of them reach term and the ones that do are associated with major congenital defects.[10] In other words, the fetal abnormality with which Jess and Steve have to deal is a consequence of a failure of the biological elimination mechanism.

In the situation in which they find themselves, these are theoretical considerations for Jess and Steve. But they serve as reminders that embryonic development involves a hazardous series of intricate events that frequently do not eventuate in a functioning new individual, let alone a healthy one. Bland as such scientific data may be, they should not be ignored in theological debate or in the decision-making of Christians. Many newly fertilized embryos will never see the light of day; they will never become the children (and subsequently adults) we may want and on whom we may place enormous theological weight (for further discussion see chapter 3). For Jess and Steve there appear to be three options.

Option 1. Continue with the pregnancy, but do not go in the direction of major surgery after birth. If this path is chosen, Jess and Steve will have to accept that their child will receive palliative care for the short time (a few days at most) he or she will be alive. They will be accepting that they are not going to do 'everything' possible for the child in the way of surgical interventions. They will be accepting the reality of death, but will be doing this in the knowledge that the child will be cared for and loved during his or her short life.

---

9   J. J. Schlesselman, 'How Does One Assess the Risk of Abnormalities from Human in Vitro Fertilisation?', *American Journal of Obstetrics and Gynecology* 135 (1979), 135–48.

10   D. Gareth Jones, *Bioethics: When the Challenges of Life Become Too Difficult* (Adelaide: ATF Press, 2007).

Option 2. Continue with the pregnancy with the prospect that major surgery will be commenced almost immediately after birth and will be followed by repeated surgical interventions. They will be doing everything possible for the child by way of utilizing the technology available to paediatric surgery. They will do this in the expectation and hope that their child will be with them for a few years, even though they also know that in all probability their child will be very unwell and will spend a considerable amount of time in hospital. They will be utilizing the most sophisticated array of technological gadgetry and expertise currently available to rectify congenital malformations, and they may regard this as their duty towards their child, for whom they have a God-given responsibility.

Option 3. They opt to terminate the pregnancy. They know that the outcome will be bleak whatever course of action they adopt. For them there is no easy road, and while they would not normally go in the direction of abortion, they see this as the quickest way to solve a dilemma they feel is beyond them. This will be no light decision on their part, and they know it will be accompanied by deep regrets, but for them it may be a tenable way forward.

This brief overview of the three options facing a couple in this type of situation demonstrates that there is no guilt-free answer. They are in a heart-rending place where their hopes and aspirations have been dashed no matter which option they decide upon. Any one of the possibilities open to them will seem to be tainted by ethical (and for Christians theological) compromise; each is accompanied by competing arguments. These comments apply to any couple in this position, but the following discussion applies particularly to a Christian couple.

The context for a decision in this instance is their strong Christian faith, and the enormous support they know they will receive from family and church community. However, what if the situation had been quite different: they had only recently moved from another country and so had no family around them? Or to make matters more demanding, what if their new church community was not supportive since it was prepared to consider only one way forward – heroic surgical measures to save the life of the child?

Each of the options raises matters that extend beyond this specific case, although they are brought to a pinnacle here. I am looking at it in the context of reproductive biology and bringing new life into existence. I have deliberately veered from the usual territory of IVF and genetic issues (although I return to these in chapter 3), since this less well-trodden area brings to light surprisingly difficult demands at the interface of clinical science and faith.

In this instance, any technological prowess is not directed at bringing new life into existence; that has occurred using natural means. What is at stake here is whether a natural process that has gone wrong can be redirected using technology. To what extent should technology be employed to rectify abnormalities? So often this question is asked later in life when dealing with adults, and then the adults themselves are in a position to make their own decision (even when a hard one). In this case the technology is needed to enable the child to grow and develop. This may be relatively unproblematic if there is reasonable assurance that it will work well and that it will be successful, but that is not the case here. There will be failure in the sense that the child will not live beyond a very few years, and those years of life will be severely compromised and perhaps debilitating for the child.

Option 1 is the path of acceptance. I use this in a positive sense, in that it is acceptance knowing that technology will not bring about the sort of result that would be welcomed. It is not the result that would be for the good of the child. This shows that the child's interests are being taken into account as well as the parents' interests. Of course, this involves a particular judgement, that the child's death (non-existence) is preferable to a few years of limited and possibly illness-plagued existence. But it is made on the basis of the best clinical evidence available, and in light of considerable reflection and spiritual direction. Within a Christian perspective it reflects an unwillingness to abort the fetus, and a view that the child is in God's good hands no matter how brief its existence outside the womb. It accepts that God has purposes for the child and parents even in the midst of sadness and tragedy. It does not point to a rejection of technology, since that would have been utilized if there had been evidence that the end result would have been what the parents would regard as a beneficial outcome. It signals acceptance of the child, even in the face of the overwhelming odds against it. By continuing with the pregnancy the mother is able to bond with the developing individual, and both parents have an opportunity to grieve at the loss of a loved one, who is a member of their family. When confronted by profound loss, learning to grieve in the midst of acceptance is an expression of love.

Option 2 is the path of striving, in the sense of resorting to the most that technology is able to offer. As in many other areas of life this is the conventional and readily accepted path. If the technology is available, use it. Most people, including Christians, do this repeatedly. Concerns only tend to arise when there are resource implications (it is very expensive and/ or it is not being fairly allocated), it compromises other worthwhile goals (by diverting our attention or wasting our energy), it has negative side effects and it fails to live up to its claims. In this instance, the technology will not achieve what anyone would like it to achieve, and yet it will give a few years of life where otherwise there would be no life. This is a paradox, since the technology will have a benefit – that of giving life – but this may be seen as a dubious benefit. Nevertheless, the attraction for some will be palpable; they will be doing something to help their child (and perhaps themselves). In Christian terms this may be interpreted as accepting the life that God has bestowed upon this child, a child who is in the image of God. By doing everything possible to assist in giving life and in celebrating life, the parents may see themselves as acting with God and fulfilling his life-giving purposes. Similar drives are seen when parents refuse to accept that conventional medical therapy has failed to prolong the life of a child, and they resort to unconventional therapies, often at great expense and in the absence of reputable scientific evidence for their efficacy.[11] For them, this is God's will; they are doing everything for their child. While one has great sympathy for the parents' predicament, one has to question whether they are not in some of these instances demonstrating an undue reliance on the technological imperative and human ability. Are they even 'playing God', a criticism usually levelled at bringing life into existence or in ending life?

11   Jennifer Jackson, 'Unproven Treatment in Childhood Oncology – How Far Should Paediatricians Co-operate?', *Journal of Medical Ethics* 20/2 (1994), 77–9.

Option 3, resorting to abortion, is the path of acquiescence. While a decision involving abortion will inevitably revolve around the merits or otherwise of induced abortion, in this case it would be aborting the fetus in the face of inevitable death. The life of the fetus and subsequent child will be short, especially so if option 1 is contemplated. What then is the virtue of continuing beyond the time of diagnosis of this massive abnormality? For those to whom abortion cannot be contemplated under any circumstances there will be no choice. But if this is not the case, why not go down the road of abortion in this particular set of circumstances? The answer given will depend upon a number of factors. What is the parents' view of the fetus and future child, a child originally wanted? What are the priorities of the parents and the convenience of the mother, who does not have to continue with a pregnancy, go through labour, and explain to every inquiring person that the baby is not going to survive birth? The conflict in this instance is between the longing for a new life and knowing that that life will never eventuate beyond a very brief period of time. Compromise is inevitable no matter what form it takes. Ultimately, though, abortion signals rejection of the child, no matter how understandable the predicament for the mother. While I do not believe this option should always be condemned, I hesitate to advocate it as the preferable way forward. It is taking life as against accepting death (option 1), a distinction that some will see as crucial in signifying the nature of one's relationship to God and his purposes.

In each of these responses the context within which the couple is living may play a significant role in the final outcome. Considerable support from family and community is required in each of these options, although if this is lacking option 3 may appear the preferable one for some. This is not to suggest that this option should be adopted, but to emphasize that decisions of this order should not be individual ones. The environment is critical and couples like this require as much human support as possible. The context within which a final decision is taken may be decisive in determining the final outcome, and this is where fundamental values, including Christian ones, will come in to play.

Neuroscience: Altering the functioning of an existing life

Two possibilities will be considered in the neuroscientific domain. Condition 1 is a pathological state and attempts to remedy it; this is the world of therapy. Condition 2 takes intrusion further to investigate whether relatively normal brain functioning should ever be improved; this is the world of enhancement. The contrast with the reproductive domain is obvious, and yet in both areas the attempt is to change people using the tools of science, and the question is whether such tools can be employed to bring glory to God. With condition 1 all agree that something has gone wrong – there have been pathological changes to the brain. And yet the same was true with both reproductive cases investigated, both of which were pathological in nature. Do we try and rectify these pathological states, and if so, how do we do so? And what theological guidance do we have? Condition 2 enters new territory, with its ethos of improvement, but we constantly strive to enhance our prospects and ourselves, and so it is not the notion of enhancement that is problematic. The stumbling block appears to come with the manner in which the tools of science would be employed.

Condition 1 – behavioural abnormalities

The first example is that of very severe depression. This is the real life example of Cathy Wield, a Christian who suffered extreme anguish due to depression. In telling her story at the end of her turmoil, she wrote, 'My life journey took me through a single, but continuous, seven-year episode. It was a terrible nightmare of torture and imprisonment. I am one of the fortunate ones to have survived and recovered'.[12]

She was a junior doctor who suddenly found that she felt very unwell and was unable to cope with the demands made upon her both

---

12   Cathy Wield, *Life after Darkness: A Doctor's Journey through Severe Depression* (Oxford: Radcliffe, 2006), xv.

professionally and in her personal life (she had four young children). The depression from which she was suffering turned out to be profound, long-term and debilitating. As a result her life and inevitably that of her family turned out to be catastrophic for the next seven years. Over this extended period she suffered from the most extreme mental anguish that proved resistant to a wide array of routine medical treatment, while the support of family, church and community failed to alleviate it in any discernible way. Only repeated bouts of electro shock therapy (ECT) made any difference, but the improvement was always temporary. Over this horrific seven-year period she was admitted to five different hospitals, was under the care of seven consultant psychiatrists, and was administered thirteen different classes of drugs. She also had psychotherapy sessions at least once a week. For two out of the four years she was in hospital she was on enforced detention, since she was a danger to herself having attempted suicide on a number of occasions.

Five years after she was first diagnosed with severe depression, Cathy Wield was referred for neurosurgery for mental disorder (NMD). This is the modern version of what was classically referred to as psychosurgery, a procedure that fell into disrepute in the 1960s and 1970s. While there are similarities between the two, since they both seek to alter behaviour by carrying out surgery on the brain, the nature of the lesions made is quite different. Currently, discrete lesions are made with precision in one small brain region (the location of which differs depending on the rationale for the surgery). In Cathy's case the surgery took place eighteen months after referral for NMD in 2001. Eight days after the operation Cathy's depression suddenly lifted.

> I believed that God had stepped in and his healing touch had been upon me. No one could give me any medical explanation for what had happened, but whatever anyone believes, there is no doubt that something truly remarkable had put me back into life without the need for any of the extensive rehabilitation programme which had been planned.[13]

13   Wield, *Life after Darkness*, 185.

The ethical and clinical issues raised by NMD are considerable and this is not the place to debate these.[14] For my present purposes, I am accepting the validity of using NMD under very specific conditions and in very well-defined circumstances such as this one. In no way is it a panacea for severe depression or other behavioural conditions like extreme aggression, sexual problems, or obsessive-compulsive states. Any physical intrusion into the brain, whether surgical as in this instance, or pharmacological, has to be carried out with great care, since it is the centre of what makes us what we are.

In Wield's case, the rapid recovery was completely unexpected. The recuperative period and gradual return to normality that was anticipated was bypassed. Writing seven years after her initial recovery, she has had one relapse and is not drug free.[15] While she is now retired from work as a psychiatrist, she can function in society and is an advocate for greater understanding of depression, self-harm and suicide. In particular, she wants to see a reduction in the stigma associated with mental illness, and a far deeper understanding in the church.

While Wield's recovery and ongoing functioning cannot be readily explained, either neurologically or theologically, it illustrates the point that alterations to the brain can have major consequences for good (as well as bad) in an individual's mental state. There can be no escape from the fact that mental and behavioural problems stem from pathological occurrences in the brain, resulting in potentially devastating consequences for the integrity and wholeness of a person. It also demonstrates that, extreme as this may be, on occasion surgery on the brain may improve severely abnormal behaviour.

This is an extreme case and yet it is not unique. The challenge it poses from a Christian angle is what authority any one has to alter an individual's brain in this manner. When dealing with an existing person, we are

14   See, for example, Raj Persaud, David Crossley, and Chris Freeman, 'Should Neurosurgery for Mental Disorder Be Allowed to Die Out?', *The British Journal of Psychiatry* 183/3 (2003), 195–6.

15   Cathy Wield, *A Thorn in My Mind: Mental Illness, Stigma and the Church* (Watford: Instant Apostle, 2012).

dependent upon the wishes and consent of the person in question, and yet in a situation such as this one, that person's brain and behavioural responses are seriously maladapted. One has to question whether there are substantial differences between this and altering genes, or the way in which genes are expressed in embryos or adults? There can be little doubt that there are differences between the embryonic and adult cases, but these have nothing to do with genes. Additionally, it is crucial to determine what constitutes normal behaviour. Debilitating depression is far from normal, and yet less severe depression and difficult-to-control violence occupy far more borderline territory.

Neurosurgery for mental disorder

In neurosurgery for mental disorder very small relatively discrete areas within the brain are destroyed by passing electric currents through implanted electrodes.[16] One of the best attested operations today is cingulotomy, an operation in which a bundle of nerve fibres connecting the frontal lobes with the limbic system is interrupted by precise lesions. The principal indications for cingulotomy are depression and obsessive-compulsive states.[17] The other major approach is a lesion in a small region known as the amygdala, and this is carried out to control extreme violence and aggression.[18]

---

16   David Christmas et al., 'Neurosurgery for Mental Disorder', *Advances in Psychiatric Treatment* 10/3 (2004), 189–99; Keith Matthews and Muftah S. Eljamel, 'Status of Neurosurgery for Mental Disorder in Scotland: Selective Literature Review and Overview of Current Clinical Activity', *The British Journal of Psychiatry* 182/5 (2003), 404–11.

17   Hyun Ho Jung et al., 'Bilateral Anterior Cingulotomy for Refractory Obsessive-Compulsive Disorder: Long-Term Follow-up Results', *Stereotactic and Functional Neurosurgery* 84/4 (2006), 184–9; J. Douglas Steele et al., 'Anterior Cingulotomy for Major Depression: Clinical Outcome and Relationship to Lesion Characteristics', *Biological Psychiatry* 63/7 (2008), 670–7.

18   Maria Mpakopoulou et al., 'Stereotactic Amygdalotomy in the Management of Severe Aggressive Behavioral Disorders', *Neurosurgical Focus* 25/1 (2008), E6.

Operations such as these are only carried out as a last resort. The situation is never simple, and enormous care has to be taken before lesioning a part of the brain for behavioural reasons (as opposed to the removal of a tumour or treating a hemorrhage). The complexity of structure, the intricate network of connections, coupled with the relatively primitive state of knowledge about the brain, mean that all such operations will be controversial.

But can they be justified in Christian terms? What does it mean to modify the way in which someone's brain functions, even if it appears to be functioning abnormally? This raises the question of whether one human being has authority to do this to another human being – not using ethical or medical criteria, but theological ones. No matter what the intentions or even the end-result, this is modification to an individual's inner being – some would say 'soul'. We know that such modification occurs when things go wrong, whether the result of brain injury, tumours or hemorrhages (see chapter 5), but this may not justify modifying someone's brain on the ground that we (and others) dislike how that individual is behaving. We may consider that the behaviour is so difficult and unpleasant that any restraint is justified and that the ends are good ones, and yet the means used are draconian. They are permanent and may have side effects.[19]

We are getting close to changing what an individual is as a person, possibly affecting their response to, and approach to, God. But in this instance we seem to be embarking on these activities for very serious and eminently worthy reasons, as we stand alongside the individual in their misery and weakness. There is no hint that we are doing it for our own self-aggrandizement, nor are we trying to improve the individual in some superficial manner.

There are numerous ways of doing this: behaviourally, electrically, chemically, surgically, and through counselling. The effects of some of these are temporary; in other cases they are permanent. The alterations, of whatever kind they may be, are to existing individuals, who, along with the network of relationships and interactions that constitutes their community,

19   Matthews and Eljamel, 'Status of Neurosurgery for Mental Disorder in Scotland'.

are changed. Unlike the debate in the reproductive area, we are no longer dealing with potential or prospective individuals, but with individuals with their own history and trajectory.

We may be justified in acting in these ways, as long as the bar for doing so is set very high, and the goal is always that of benefitting the patient. As far as possible fully informed consent is obtained from the patient, although this may be difficult to obtain if their ability to understand the consequences of what is proposed is compromised. As humans made in the image of God we are creative beings who are to use this creativity to assist others reflect their own God-like attributes as effectively as possible. We are to do our best to overcome debilitating mental and brain pathologies, realizing that this may involve demanding levels of clinical judgement, pastoral care and support, and spiritual counsel. But this simply reflects our role as stewards of God's creation.[20]

In this realm the issue is not that of uninhibited technological imperialism. Nevertheless, the technology is impressive (or threatening depending upon one's perspective) and any decision-making will hopefully be informed by value systems that elevate the status and worth of people. There is nothing inevitable about this, but that is hardly unusual. This is why using a case study like that of Cathy Wield is so helpful; here was someone in dire need of assistance, someone who was not hankering after idealistic ends, but after healing. The end in sight was the alleviation of mental distress and suicidal intentions. Christian compassion cannot overlook the desperate need of someone in this state, and if a particular technological approach can contribute towards a positive therapeutic outcome, the response should be one of gratitude rather than condemnation. But this is only made possible by an attitude that is prepared to accept a technological contribution, even a controversial one.

---

20   Allen Verhey, "'Playing God" and Invoking a Perspective', *Journal of Medicine and Philosophy* 20/4 (1995), 347–64.

Condition 2 – cognitive enhancement

> Felicity is an ambitious student and is determined to get into the university of her choice. She has always worked hard, but knows that the competition will be cut-throat. She also knows that the students around her are taking methylphenidate (Ritalin) in order to stay awake and alert for much longer than they would otherwise be able to do. Realizing that she may be disadvantaged if she doesn't follow suit, she starts using this non-prescription drug.
>
> She also knows that it acts directly on her brain by increasing dopamine levels, and that it could have side effects on other organs. She is dimly aware of the ethical issues raised by taking this drug, but is more conversant with its abuse potential and questionable legal status. She has not thought at all that what she is indulging in is a form of neurocognitive enhancement.

With this scenario we move from therapy to enhancing a life, or to what is commonly perceived as enhancing a particular function. The possibility of enhancing the lives of ordinary people through biomedical technology is present reality. What we encounter here is the use of common medications for purposes for which they are neither prescribed nor marketed. Examples abound as drugs originally designed to treat a medical condition are employed by healthy individuals in attempts to improve their performance. The illustration given here is just one example, since Ritalin, which has conventionally been prescribed for children suffering from ADHD (attention-deficit hyperactivity disorder), and Provigil (Modafinil), prescribed for individuals with narcolepsy, also appear to be useful in aiding concentration, alertness, focus, short-term memory and wakefulness in healthy individuals.[21] Yet another example is Donepezil (Aricept) originally developed as a treatment for Alzheimer's disease, which has been shown

---

21   R. Elliott et al., 'Effects of Methylphenidate on Spatial Working Memory and Planning in Healthy Young Adults', *Psychopharmacology* 131/2 (1997), 196–206; M. A. Mehta et al., 'Methylphenidate Enhances Working Memory by Modulating Discrete Frontal and Parietal Lobe Regions in the Human Brain', *Journal of Neuroscience* 20/6 (2000), RC65; D. C. Turner et al., 'Cognitive Enhancing Effects of Modafinil in Healthy Volunteers', *Psychopharmacology* 165/3 (2003), 260–9.

to improve recall of training when taken by healthy, but older, pilots in a flight simulator.[22]

The use of prescription drugs by healthy adults as forms of cognitive enhancement marks a shift of major proportions from even NMD as in the previous section. And yet the shift may be far from apparent, since these drugs are being taken to strengthen what appear to be culturally desirable traits, such as improved productivity and heightened focus.[23] Use of drugs to act as stimulants – from energy drinks to amphetamines – is widespread and these cognitive neuroenhancers simply extend the list. However, as outlined in chapter 5, they come with problems, chief among which is addiction. They also serve as a technological fix, a short cut that may or may not have the desired effect, and may be accompanied by long-term consequences. If they serve as a substitute for adequate sleep, good nutrition and sensible study habits, their enhancement function may be deceptive.

In spite of these provisos it would be unwise to refuse to face up to the challenges posed by these and other forms of enhancement. The borderline between therapy and enhancement can be exceedingly difficult to determine, with procedures ranging from what most would consider to be obviously beneficial through to the highly questionable, and from the uncontentiously therapeutic to what in the eyes of many would be flippant and superficial.[24] However, the slide from undoubted therapy to dubious enhancement is gradual, since we are being enhanced all the time in subtle ways. While enhancement is frequently considered to represent an improvement in human functioning beyond what is necessary for good

22   J. A. Yesavage et al., 'Donepezil and Flight Simulator Performance: Effects on Retention of Complex Skills', *Neurology* 59/1 (2002), 123–5.

23   Conan Milner, 'Study Drugs Unsafe and Unethical, Say Neurologists', Epoch Times, (27 March 2013). <http://www.theepochtimes.com/n2/united-states/study-drugs-unsafe-and-unethical-say-neurologists-369451.html> accessed 1 April 2013.

24   D. Gareth Jones, 'Enhancement: Are Ethicists Excessively Influenced by Baseless Speculations?', *Medical Humanities* 32/2 (2006), 77–81.

health, it is increasingly becoming evident that it is integral to what we now take for granted as good health.[25]

Theological assessment of enhancement has to be carried out against the backdrop of this continuum, from what is generally considered to be beneficial and praiseworthy to what may be viewed with intense suspicion. The temptation is to reject every hint of enhancement on account of the latter, and yet this cannot be justified. To adopt a precautionary stance, with its cautious emphasis and bias towards a negative assessment of unknown technologies fails to grapple with the issues theologically and is unable to critique the nuances of the debate.

We need to consider the manner in which biomedical technology is changing our conception of what constitutes the good life – or in Christian terms the life of faith. These include the day-to-day medical expectations experienced by most people in developed countries. In these countries enhancement began many years ago, as medicine and allied technologies came into their own with vastly improved public health measures, vaccination, antibiotics, antipsychotics and antidepressants, sophisticated surgical procedures and revolutions in obstetric care.[26] Even if one wishes to class some of these as therapies, the theological repercussions are the same. Individuals and whole communities now expect diseases to be conquered and even vanquished, illnesses to be cured, and life expectancy to be increased. These are legitimate Christian aspirations on condition that there is an acceptance of their limits and that they are not part of some grand scheme for reshaping the meaning and aspirations of human existence.

There is no place within Christian thinking for self-gratifying fantasies whereby human suffering and pain can be totally and finally eradicated by technological prowess. This applies as much to mental distress as to any other example of suffering. Vistas based on drug-enhanced means of drastically improving mental performance are, in all probability, deceptive

---

25   D. Gareth Jones, 'Enhancement: Is Baseless Speculation Misleading Theologians and Bioethicists?', in R. J. Elford and D. G. Jones, eds, *A Tangled Web: Medicine and Theology in Dialogue* (Bern: Peter Lang, 2009), 123–42.

26   D. Gareth Jones and Maja I. Whitaker, *Speaking for the Dead: The Human Body in Biology and Medicine*, 2nd edn (Aldershot: Ashgate, 2009), 233–4.

allurements (see chapter 6). Beyond the inevitable limitations of the science involved, the notion of radically enhancing what we are as people solely by neural manipulation stems from the view that we are nothing more than human machines. Not only does this deny a role for our interaction with others in human community, it removes us from our relationship to God and our responsibilities before God. It is, in essence, a non-theist view (an atheistic view) of human beings, according to which purely mechanical means are all that is required to improve mental functioning and transcend genetic and social endowments.

Ultimately, what place is left for God's purposes and human responsibility? This question is not asked in the sense that God is only introduced once physical processes let us down, but in the deeper sense that these processes cannot be brought to completion in the absence of an awareness of the relevance of a relationship with God. This is an acknowledgement that, although we are embedded in a world of enhancement, just as importantly this world arises from the creativity bestowed upon humans by God. We are his images who live in community; finite and also tarnished. Consequently, enhancement technologies are to be utilized in ways that promote community and co-operation rather than pride and self-absorption (1 Cor. 12:22–25). If we accept the weaker members of our communities, especially the poor, socially marginalized and the sick (Mat. 25:31–46), we will also learn to accept our own weaknesses and our own finiteness.

But is this moving too far from the illustration of students taking cognitive enhancing drugs with the aim of improving their short-term memory and their ability to concentrate? These students are not setting out to make profound statements of any description; they are simply trying to improve their examination performance. Perhaps. But Felicity, the student in the illustration, needs to be aware that getting ahead in this way is a short-term measure; others will do the same, and then they will all be looking for more effective means of getting ahead yet again. The side effects cannot be ignored, and as neuroenhancing drugs are increasingly employed, side effects will in all probability become ever more troublesome. Underlying these immediate considerations are the ones touched on already. Direct control of the brain via drugs may fail to take account of social and psychological considerations, let alone theological and ethical ones. Such considerations usually feature as part of any serious therapeutic analysis but are just as often cast aside as irrelevant when enhancement is the chosen option.

The cognitive enhancement in which students are indulging is at odds with Christian conceptions. By using a powerful means of manipulating their brains, possibly with risky and unforeseen consequences (Modafinil in particular is known to be addictive),[27] and for immediate self-centered ends, they are failing to accept the limits inherent in what they are as embodied persons. Cosmetic neurology of this ilk is misleading enhancement that fails to glorify God in a scientific milieu. Its deceptiveness stems from its excessive dependence upon scientific propensities, to the partial or complete exclusion of a God-centered perspective.

As has become all too evident in a variety of sports, when enhancement regimes (drug taking) become endemic, the pressure for the majority to follow suit is intense.[28] This is necessitated by the competitive element in professional sport, made all the more pressing by the powerful impetus of high financial stakes. What then becomes of enhancement, when all (or most) participants are enhanced? The playing field has been shifted and the expectations have been raised. All will become equal again, but this time the so-called level playing field is an inherently enhanced one. The futility of this exercise only becomes evident in hindsight, as the means of enhancement has to be continued in order to provide this new level playing field. In doing so it emerges that untoward restrictions have been unwittingly introduced.

Reflection on these two conditions (NMD and enhancement) in neuroscience pin-point crucial differences. The question: 'what will serve best a human being or human beings?' throws light on the goals of the two technological approaches. No matter how considerable the challenges of NMD, they can be directed in ways that will assist patients. By contrast, the

---

27   A. Heinz et al., 'Cognitive Neuroenhancement: False Assumptions in the Ethical Debate', *Journal of Medical Ethics* 38/6 (2012), 372–5.
28   J. Savulescu, B. Foddy, and M. Clayton, 'Why We Should Allow Performance Enhancing Drugs in Sport', *British Journal of Sports Medicine* 38/6 (2004), 666–70.

example of cognitive enhancement only assists by allegedly improving an individual's performance in a competitive environment. The enhancement is illusory, and if everyone resorts to the same means of enhancement, any anticipated advantages will disappear.

Reproduction and neuroscience

In directing attention to these four areas within the reproductive area and neuroscience, four pertinent means by which technology impinges on Christian thinking and attitudes have been covered: its use to bring new lives into existence; its use or non-use when determining what to do in the face of congenital abnormality; its therapeutic uses and its cognitive enhancing uses by existing individuals.

My argument has been that intrusions in all these areas represent significant control by human beings, but that from a Christian perspective all represent potential means for glorifying God and serving humanity. But by the same token all have their drawbacks. What shines through as of crucial importance is that embryo, fetus and adult are all viewed in their wholeness rather than as machines to be deconstructed. However, the very success of science stems from its power to dissect and analyze the smallest of components in their isolation – ovary, embryo/fetus, amygdala, dopamine. The tendency to lose sight of individuals in their integrity and cohesion is immense, although herein also lies the ability to improve people's lot by overcoming disease and debility, by improving the living conditions of individuals and communities, and even by changing aspects of human nature.

The reproductive domain is not as distinct from others as sometimes assumed. True, it has peculiar challenges and its dimensions may give the impression of being unduly far-reaching, since yet-to-be-born individuals are the targets of any interventions. But are these challenges of a different order to those raised by profound changes to existing individuals? I suggest that they are not, and that criteria used to guide decision-making in one area also apply in any other ones. From a Christian perspective we always need to ask pertinent questions:

- Will use of the procedure help those affected to better image God and acknowledge him as creator and sustainer?
- Will use of the procedure facilitate patients' relationship with God, or not?
- Does use of the procedure take into account human finiteness and our ultimate dependence upon God?
- Will use of the procedure enhance or detract from the fundamental role of community and family?
- Is use of the procedure aiming to bring about bodily perfection, and the elimination of all suffering?

Questions along these lines help to determine the underlying ethos within which technology is or is not being employed – at the start of life, the end of life, or anywhere in between. While I have concentrated on the individual level, these considerations also apply at the population level. In each of the questions I have inserted 'use of' since ultimately it is the way in which procedures are used that is crucial; the procedures themselves do not inherently glorify God or detract from him.

Whether at the start of a new life or when dealing with an ongoing life, the Christian vocation is to wait patiently for the redemption of our bodies (Rom. 8:23). Simultaneously, we are to seek to improve lives in ways that will assist others in seeing their place within a world that is dependent upon God while thankfully making use of the abilities and gifts he has so richly bestowed upon us. Being faithful stewards of God in a scientific milieu is not without its challenges and highly controversial aspects, at least as much in neuroscience as in reproduction.

Artificial reproductive technologies and pre-birth dilemmas

Delight and disquiet

Reproductive issues bring us face to face with quite profound theological questions, as profound as anything encountered in more general biological realms, or in any of the physical sciences. The reason is that the reproductive technologies confront us in very practical terms with the relationship between human and divine control, between that which should or should not be subject to the interference of human beings, and what should or should not be left to the whims of fate/chance or the dictates of God. In other words, they raise the question of where God enters the picture, if indeed there is any place for God or any god.

It is difficult to overestimate the seismic effects that IVF (*in vitro* fertilization) has had on all modern societies. Its use in bypassing infertility has been overshadowed by the many other uses to which it can be put in transforming the reproductive process, and in contributing to medical progress. These and other consequences of IVF stem from the ability to investigate and manipulate human embryos in the laboratory. Consequently, the human embryo has lost its mysterious and unfathomable quality, and this is the root of the disquiet expressed by many when confronted by techniques for controlling embryonic development.

Sir Robert Edwards, the Cambridge physiologist whose ground breaking work in the 1960s and 1970s led to this revolution, had to wait until 2010 to be awarded the Nobel Prize for Physiology or Medicine. However, even that event was contentious, with those who applauded it matched by voices raised against it on the ground of its untoward social and ethical repercussions.

Monsignor Carrasco, the Vatican's spokesman on bioethics, speaking in a personal capacity, considered that the award was 'completely out of order,' since it ignored the ethical questions raised by this form of fertility treatment. Without this treatment there would be no market for human eggs, and there would not be numerous freezers filled with surplus embryos.[1]

The numerous photos of happy families with their smiling and healthy IVF children are matched by the dire prognostications of doomsayers bemoaning its very existence. For the former some of the limitations of human nature have been overcome; for the latter human nature has been blighted and torn asunder. These vastly discrepant responses originate in different reactions to the ability to interfere with, and direct, embryonic development. Those who emphasize its legitimacy accept the increased control now available to humans; those who regard it as unacceptable fear where the resulting hubris will lead, on the ground that control of our earliest beginnings signifies control of our lives as human beings. All fears stemming from this second form of control are expressed as opposition to the first type of control.

Theologically, the concern is that humans are intruding into what some take to be territory peculiarly set apart for God's actions. This is because the manner in which our lives commence and the way in which they end supposedly lie entirely in God's hands. In other words, there are realms that lie outside the influence, or even interest, of human beings. God brings each of us as individuals into existence, and God brings our lives as individuals to an end. To interfere in either case is to act in ways that challenge God's purposes and plans. But if this is even remotely true, what becomes of efforts to overcome infertility or to prolong someone's life by technological means? The former at the beginning of life poses greater theological challenges for some Christians than the latter at the end of life.

---

1   BBC News, 'Vatican Official Criticises Nobel Win for IVF Pioneer', (4 October 2010) <http://www.bbc.co.uk/news/health-11472753> accessed 17 June 2013.

A clear manifestation of these concerns at the earliest stages of human existence is provided by public debate on procedures such as IVF, PGD (pre-implantation genetic diagnosis), the production of surplus embryos in IVF programs, postmenopausal pregnancy, or research on human embryos.[2] Many of the submissions one reads in response to public consultations will be opposed to some or all of these procedures, with a sizeable number opining that scientists and clinicians are 'playing God', and that they are entering divine territory, forbidden to all but God.[3] The religious connotations are only too clear. Alongside this are concerns that any interference with human embryos will result in 'designer babies',[4] once again implying that the modification or selection of embryos is taking us into highly suspect, and even forbidden, territory.[5] Repeatedly, the message that comes across is that under no circumstances should human beings be permitted to meddle in the biological characteristics of future individuals, since design in this sense is a step too far for created beings, as though one were designing a house or car. Objections to numerous speculative possibilities lead some to reject procedures common in the clinic and well known to genetic counsellors. Since all these procedures are now integral

---

2　See, for example, Advisory Committee on Assisted Reproductive Technology, 'Use of Gametes and Embryos in Human Reproductive Research: Summary of Submissions', Ministry of Health (September 2007) <http://www.acart.health.govt.nz/moh.nsf/pagescm/6730> accessed 17 June 2013.

3　J. Stephen Bellamy, 'Evangelicals and Embryology: Responses to the Human Fertilisation and Embryology Bill', in D. Gareth Jones and R. John Elford, eds, *A Glass Darkly: Medicine and Theology in Further Dialogue* (Bern: Peter Lang, 2010), 157–90.

4　According to the Oxford English Dictionary a designer baby is 'a baby whose genetic makeup has been selected in order to eradicate a particular defect, or to ensure that a particular gene is present'. *Oxford English Dictionary*,(Oxford: Oxford University Press, 2013). This is a careful definition that steers clear of including genetic engineering and manipulation, and genetic selection for a range of speculative outcomes.

5　Jones, *Designers of the Future*. In the ACART consultation the concerns of those opposing research on human embryos was 'on the grounds that they are human life and any manipulation, including *in-vitro* fertilisation (IVF) and pre-implantation genetic diagnosis (PGD), is akin to harming or killing a person.' Advisory Committee on Assisted Reproductive Technology, 'Use of Gametes and Embryos'.

to the reproductive technologies, objections of this ilk pit these particular religious (and quasi-religious) perspectives against scientific medicine.

An underlying premise

Integral to these religious perspectives is the view that human embryos have a moral value equivalent to that of adults. The basis for this stance is the position that human life commences at fertilization: 'human life is sacred from conception'.[6] Starting from this premise there is no way in which any interference with embryos in the laboratory will be accepted. Hence, the categorical assertion that to allow any destruction of human embryos, under any circumstances, is contrary to God's ordinances.[7] For many within the Christian community this is regarded as a faithful out-working of their Christian worldview.

This stance follows from the position that this is a definitive position found in Scripture. It is true that women in the Bible are referred to as conceiving, but the writers of the time knew nothing about conception or fertilization in the way in which we do. The Hebrew and Greek phrases translated as 'conceived' ought to be rendered 'become pregnant'. It is not clear whether the unborn were thought of as persons in anything like a modern sense.[8] Of course, much was known about pregnancy, and about the growth of the unborn 'child'. This is seen for instance in Job 10:10 and Psalm 139:13–16. What we have in both cases is poetry, and so it is not possible to be sure whether the wording represents what the writers thought was actually happening inside the womb, or even what they knew. There are certainly indications that the unborn were recognized as developing

6   Congregation for the Doctrine of the Faith, 'Donum Vitae'.

7   Sacred Congregation for the Doctrine of the Faith, 'Declaration on Euthanasia', (1980) <http://www.vatican.va/roman_curia/congregations/cfaith/documents/rc_con_cfaith_doc_19800505_euthanasia_en.html> accessed 4 April 2013.

8   R. B. Hays, *The Moral Vision of the New Testament: Community, Cross, New Creation: A Contemporary Introduction to New Testament Ethics* (San Francisco: HarperSanFrancisco, 1996), 447–9.

from something formless into something developed and complete. It is also very clear that both the psalmist and Job believed that God was intimately involved in these processes.[9]

The writers in Scripture knew about miscarriages, and induced abortion was practised and punished in the ancient Near East. However, biblical law was silent on the subject. One interpretation of this is that abortion was not commonly practised in ancient Israel, and in all probability the unborn were highly valued in ancient Israel. However, the Bible presupposes a social and cultural situation totally different from that of today. Children were almost uniformly desired in ancient Hebrew and Jewish society, unlike today. This by itself does not lead to any particular conclusion on the rights or wrongs of how embryos are to be treated in multicultural and often secular societies, but the stark differences should not be overlooked.

Beyond the arguments concerning the validity or otherwise of the precise point at which human life commences (acquires value), the significance of conception as the starting point appears to be that it alone will keep in check the rampant run-away forces of modern science, and in particular the potentially dehumanizing reproductive technologies. The impression is given that we are living on a battlefield where the future of human kind is being fought over, and the sacredness of human life from conception is the one and only weapon that will prevail. If it turns out to have feet of clay, the Christian worldview crumbles. Extreme as such a position appears, it is to be found regularly in the literature. One illustration is expressed in these terms:

> Destroying an embryo is killing a human being, 'one of us'. It is wishing someone was dead – a straightforward breaking of God's law. At the very least, it is a form of hatred which flies in the face of the command to love our neighbour.[10]

---

9   John Bryant and John Searle, *Life in Our Hands* (Nottingham: Inter-Varsity Press, 2004).

10   John R. Ling, *When Does Human Life Begin?* (Newcastle upon Tyne: The Christian Institute, 2011), 15.

This 'sacredness-conception' combination has emerged as basic cultural dogma that is non-negotiable and has, in the eyes of some, become a mark of faithfulness to Christian fundamentals. And so, when confronted by the reproductive technologies, the truly Christian response in these quarters is to question whether they should be utilized at all, because once that is allowed all moral and spiritual authority has been relinquished. In these terms the legitimacy of the sacredness-conception combination is critical.

Away from these polemical debates, other Christian writers also contend that the embryo should be viewed as a neighbour, on the basis of the parable of the Good Samaritan (Luke 10:25–37), in which a foreigner took pity on and cared for a stranger in need of assistance. Using this powerful model all human beings are equally neighbours – to be protected and provided with sustenance. For writers such as Waters, Messer and Vorster,[11] embryos are not to be excluded from the category of human; they are as much our neighbours as any child or adult. Consequently, nothing can or should be done that jeopardizes their welfare.

Vorster comes to this conclusion on the basis of what he claims to be three biblical concepts: the creation of the human being as a living being with the breath of God; the creation of the living being in the image of God; and a view of the human being as a covenantal being.[12] Quite apart from the significance of these concepts in their own right, their relevance to the bioethical questions of what can or cannot be done to embryos depends upon two factors: whether the neighbourly status ascribed to the embryo applies from fertilization onwards, and also that we know what it means to treat all embryos as neighbour. These are matters of intense relevance for bioethical discussion, and yet are frequently not addressed in detail by theologians. For instance, Vorster concludes that it would be immoral to

---

11   Messer, *Respecting Life*; J. M. Vorster, 'A Christian Ethical Perspective on the Moral Status of the Human Embryo', *Acta Theologica* 31/1 (2011), 189–204; B. Waters, 'Does the Human Embryo Have a Moral Status?', in B. Waters and R. Cole-Turner, eds, *God and the Embryo: Religious Voices on Stem Cells and Cloning* (Washington, DC: Georgetown University Press, 2003), 67–76.

12   Vorster, 'A Christian Ethical Perspective'.

create an embryo with the intention of destroying it in stem cell research.[13] This leaves unanswered the question of whether surplus embryos in IVF programs can be used for research, including the derivation of stem cells, or whether it is ethical to produce large numbers of embryos routinely in fertility treatment involving IVF.

What criteria do we have for determining if any or none of these actions are akin to treating embryos as neighbour? Waters concedes there may be a hierarchy of duties and obligations to different types of neighbours, and that on rare occasions 'certain neighbours (such as weak and vulnerable embryos) may be sacrificed because of the genuine needs of other (ill and injured) neighbours.'[14] However, more work is required to enunciate more precisely what these criteria are, and until this is done it is difficult to determine what role the embryo-neighbour approach might have in assessing the reproductive technologies.

Reproductive technologies as alien territory

My point in alluding to this issue is not to argue for the rightness or wrongness of this particular position but to suggest that the religious convictions of some are diametrically opposed to a whole area of scientific investigation and to the clinical practice that emanates from it. Consequently, a whole range of approaches within the reproductive technologies becomes alien territory to those with this particular religious stance. This has important repercussions.

The *first* is that those with this religious persuasion have little way of influencing public policy on the reproductive technologies, unless they can persuade policy makers that the whole reproductive endeavour on which we have embarked is massively misguided. This is the position espoused by the Vatican as reiterated in the 2008 document *Dignitas Personae*.[15]

---

13 Vorster, 'A Christian Ethical Perspective'.
14 Waters, 'God and the Embryo', 75.
15 Congregation for the Doctrine of the Faith, 'Dignitas Personae'.

One may have no qualms about this and yet it appears that the possibility of constructive dialogue has been lost, and that the possibility of counterbalancing some of the more extreme secular trends within society has been forfeited.

The *second* repercussion is that it allows little opportunity for ongoing debate within Christian circles, since Christians differ on the ethical and theological significance of fertilization. Such debate becomes superfluous if no reproductive technologies are allowable on allegedly theological grounds. While there may be considerable inconsistency in the outworking of this stance, even its theoretical formulation amounts to a stifling of any science-religion debate in this area, with negative consequences for theological input into scientific, policy and ethical decision-making.

The issue in this instance is that one particular position, that for some has theological warrant, is implacably opposed to the dominant scientific position. This does not presuppose which one is correct, but it unequivocally demonstrates that there is incongruence between the two. This is a particularly modern expression of the unease with which the intersection between science and religion may sometimes be viewed. One fascinating characteristic of this issue is its intensely practical façade. There is no way in which it can be quietly parked in the halls of academia. Those who encounter this daily are ordinary people confronted by one or other of the many facets of infertility. Rarely are they well informed about the abstruse theological arguments or complex scientific developments that surround the embryo, any more than are ordinary Christian ministers or even most theologians. The modernity of these questions is as striking as are any emotive responses.

For some, the ability to modify what human beings are under normal circumstances, have become challenges to the authenticity and authority of God. I shall argue that this is not necessarily the case, but it is a well-entrenched position that requires careful analysis. In this chapter the focus is on the earliest stages of existence, but its more extensive application in other areas of human life will appear in subsequent chapters when changes to genetic constitution, the brain and the body are assessed.

The Artificial Reproductive Technologies (ARTs) in historical context

In 1987 I wrote a book entitled: *Manufacturing Humans: The Challenge of the New Reproductive Technologies*, in the preface to which I commented that the rapid developments in the reproductive technologies were proving as frenetic in ethical and theological circles as in laboratories and clinics.[16] Little did I know at that time that such a sentiment would be even more apposite in the twenty-first century.

The debates have intensified as scientific developments in biology and medicine continue to far outstrip the ability of our ethical (let alone our theological) systems to cope with them. Modern medicine ushers in immense hope but as previously discussed, never far removed from the hopes and anticipations are perils of many descriptions. Our view of ourselves as human beings has been fundamentally changed by numerous improvements in the quality of our lives, from the earliest stages in development through to old age. In the majority of developed countries people are expected to live into the seventies and eighties as a matter of course, and it is anticipated that the wombs of the infertile will be 'opened' by IVF rather than by God. Little leeway is left to nature or natural (divine?) forces, and when these overwhelm populations or individuals, the blame is placed on God, even if the consequences are the result of human corruption or sheer bad planning. The ARTs of course are only one component of these trends and yet they feature prominently in them. To appreciate the speed of developments in the ARTs one has only to glance at the following milestones.

By 1989 more than 400,000 children had been born using IVF; that figure has now risen to 5 million.[17] In some societies 3–4 per cent (4.6 per

16 Jones, *Manufacturing Humans*.
17 European Society of Human Reproduction and Embryology, 'The World's Number of IVF and ICSI Babies Has Now Reached a Calculated Total of 5 Million', (2 July

cent in Denmark) of all children are born using IVF.[18] As I look out at a class of 200 students I would not be surprised if four or five were born using IVF, and more using donated gametes. And the same probably applies to church congregations.

In 1987 IVF was a relatively simple procedure, and few of its associated developments were even on the horizon. In particular, the technique of intracytoplasmic sperm injection (ICSI) that has revolutionized the efficacy of IVF lay in the future. Today this is involved in almost 70 per cent of IVF procedures.[19] Not only this IVF is being employed increasingly to enable older women to have children, especially those in their late thirties and forties, quite apart from the still small number of post-menopausal women. This is part of a vast transformation in social expectations that commenced with the oral contraceptive.

As increasing numbers of children have been born via IVF their health status has come under the microscope. There is evidence of a slightly higher rate of chromosomal abnormalities in children born as a result of ICSI. Studies suggest that IVF and ICSI are associated with a higher risk of rare disorders associated with the 'imprinting' of genes, occurring in approximately 1 in 4,000 children compared to 1 in 10,000 – 1 in 15,000 children conceived naturally.[20] Otherwise not too much of note has emerged to date. Women born using IVF are able to have their own children naturally.

The procedure of PGD was unheard of in 1987. With its advent in 1989 the world of genetic diagnosis was turned upside down through the newly found ability to select embryos.[21] This constitutes the basis of genetic

---

2012) <http://www.eshre.eu/ESHRE/English/Press-Room/Press-Releases/Press-releases-2012/5-million-babies/page.aspx/1606> accessed 18 June 2013.

18   A. P. Ferraretti et al., 'Assisted Reproductive Technology in Europe, 2008: Results Generated from European Registers by ESHRE', *Human Reproduction* 27/9 (2012), 2571–84.

19   Ferraretti et al., 'Assisted Reproductive Technology in Europe, 2008'.

20   Somjate Manipalviratn, Alan DeCherney, and James Segars, 'Imprinting Disorders and Assisted Reproductive Technology', *Fertility and Sterility* 91/2 (2009), 305–15.

21   Charles Coutelle et al., 'Genetic Analysis of DNA from Single Human Oocytes: A Model for Preimplantation Diagnosis of Cystic Fibrosis', *BMJ* 299/6690 (1989), 22.

selection, especially with single gene disorders, although it can be used more widely than this. In some ethical and theological literature it has given rise to concerns regarding eugenics,[22] although these concerns by themselves do not amount to anything resembling classic eugenics of the first half of the twentieth century.

Writing in 1987 there was no way in which embryonic stem cells could have been mentioned, simply because the general public had never heard of them, and no one would have considered that their utilization might have theological overtones. All that changed in 1998, when they were first successfully derived from human embryos.[23] In some quarters the ethical issues they raise eclipse even those associated with abortion, and there has been a tendency for them to dominate debate on the ARTs.

In 1987 there was little detailed discussion of any form of cloning, although I did touch on it in my writings at that time. However, it elicited little public or theological interest, and the work of developmental biologists was ignored. All this changed with the 1997 announcement of the birth of Dolly, with its avalanche of dire warnings predicting the end of humanity as it is today, with prognostications that science is out of control and has become a means of mass destruction.[24]

Not surprisingly, in 1987 practically no one knew anything about therapeutic or research cloning. Today it is recognized as a major revolution in regenerative medicine, although it has come in for a great deal of

---

22 Hamish Anderson, 'Preimplantation Genetic Diagnosis: From Clinic to Eugenic Fears and Disability Concerns', (Thesis, Bachelor of Medical Science with Honours, University of Otago, 2012) <http://hdl.handle.net/10523/2335> accessed 17 June 2013; D. S. King, 'Preimplantation Genetic Diagnosis and the "New" Eugenics', *Journal of Medical Ethics* 25/2 (1999), 176–82; Christine Rosen, 'Eugenics – Sacred and Profane', *New Atlantis* 2 (2003), 79–89.

23 J. A. Thomson et al., 'Embryonic Stem Cell Lines Derived from Human Blastocysts', *Science* 282/5391 (1998), 1145–7.

24 John Arlidge, 'Scientists "Able to Create Human Clone"', *The Guardian* (26 February 1997), 6; Tim Radford, 'German Fury over Cloning', *The Guardian* (28 February 1997), 1; I. Wilmut et al., 'Viable Offspring Derived from Fetal and Adult Mammalian Cells', *Nature* 385/6619 (1997), 810–13.

criticism on the grounds that it is a threat to human dignity.[25] Concerns
of this nature resurface whenever any new development appears in the
literature, as with the 2013 report of the success by Mitalipov and group
in deriving human embryonic stem cells by somatic cell nuclear transfer.[26]
The details of the paper and its intent soon gave way to cries of alarm at
what might befall humanity. For instance, Boston's Cardinal O'Malley
condemned the work on the grounds that it created human lives simply
to destroy them, and also that those who want to produce cloned children
as copies of other people will follow on from this work.[27]

In 1987, chimeras and hybrids were nothing more than threatening
part-human, part-animal grotesque monsters encountered in Greek mythology and science fiction. We could not have imagined that by the end of the
first decade of the twenty-first century they would have entered the domain
of serious science and their creation would be allowed in some countries.[28]
With this debate new terms entered our vocabulary, namely, inter-species
embryos,[29] or human admixed embryos.[30] While the research side of these
developments has to date proved illusory, the objections from Christian
groups have been vociferous, including claims that they constitute a threat
to the whole concept of human identity, human dignity, and human rights

25   See also John Harris, 'Cloning and Human Dignity', *Cambridge Quarterly of
      Healthcare Ethics* 7/2 (1998), 163–7; Axel Kahn, 'Clone Mammals ... Clone Man?',
      Nature 386 (1997), 119; President's Council on Bioethics, *Human Cloning and Human
      Dignity: An Ethical Inquiry* (New York, NY: Public Affairs, 2002).

26   Masahito Tachibana et al., 'Human Embryonic Stem Cells Derived by Somatic Cell
      Nuclear Transfer', *Cell* 153/6 (2013), 1228–38.

27   Kevin Clarke, 'O'Malley Condemns Cloning Breakthrough', *America: The National
      Catholic Review* (15 May 2013) <http://americamagazine.org/content/all-things/
      omalley-condemns-cloning-breakthrough> accessed 17 June 2013.

28   Calum MacKellar and David A. Jones, eds, *Chimera's Children: Ethical, Philosophical
      and Religious Perspectives on Human–Nonhuman Combinations* (London:
      Continuum, 2012).

29   Zeki Beyhan, Amy E. Iager, and Jose B. Cibelli, 'Interspecies Nuclear Transfer:
      Implications for Embryonic Stem Cell Biology', *Cell Stem Cell* 1/5 (2007), 502–12.

30   'Human Fertilisation and Embryology Act', (2008) <http://www.legislation.gov.
      uk/ukpga/2008/22/contents> accessed 18 June 2013.

by blurring the boundaries of nature, transgressing the 'kinds' of Genesis and marring the image of God.[31]

Developments in care of those born prematurely have been viewed far more positively, in spite of the very high technology involved and the mixed results clinically. In the 1980s around 8 per cent of babies born prematurely at twenty-three weeks survived. In the UK today about 50 per cent of these may survive with aggressive treatment. In the 1980s around 40–45 per cent of babies born before twenty-eight weeks survived; the comparable figure today is 80 per cent. Morbidity rates have also improved but remain a substantial obstacle to the quality of life in some of the particularly premature.[32]

Responses to the ARTs in the mid-1980s

Debate in the UK was strongly influenced initially by the original Warnock Report, published in 1984 and that set the tone for a good deal of subsequent decision-making.[33] The Inquiry agreed that human embryos have a special status and should be afforded some protection in law, but that this status is not equal to that of actual persons and so this protection may be waived for early embryos and research on human embryos allowed under licence up to fourteen days after fertilization. The Inquiry concluded that infertility is a condition meriting treatment, and that IVF and donor insemination could now be considered an established form of treatment. The Inquiry also concluded that egg and embryo donation was acceptable, with provisos.

31 Neville Cobbe, 'Cross-Species Chimeras: Exploring a Possible Christian Perspective', *Zygon* 42/3 (2007), 599–628; Messer, 'Christian Engagement with Public Bioethics'; Messer, *Respecting Life*.
32 T. Markestad et al., 'Early Death, Morbidity, and Need of Treatment among Extremely Premature Infants', *Pediatrics* 115/5 (2005), 1289–98.
33 Mary Warnock, *Report of the Committee of Inquiry into Human Fertilization and Embryology* (London: HMSO, 1984).

At that time some of the most damning condemnations of IVF and its associated procedures came from theologians. For instance, Oliver O'Donovan, who at that time was Professor of Moral Theology at Oxford University, wrote a little book in 1984 with the title *Begotten or Made?* This has become essential reading for all who express theologically grounded concerns at the ARTs and other technological interventions in selected areas of medicine.[34] He wrote, 'When we start making human beings we necessarily stop loving them; [...] that which is made rather than begotten becomes something that we have at our disposal, not someone with whom we can engage in brotherly fellowship [...] I do not know how to think of an IVF child except [...] as the creature of the doctors who assisted at her conception.'[35]

Also in 1984 Thomas Torrance, formerly Professor of Church Dogmatics at the University of Edinburgh, wrote a short booklet that was one of the most impassioned condemnations of the Warnock Report to appear. According to him, 'What is at stake is nothing less than the future of the human race, but what is also at stake is the integrity of the scientific and moral conscience [...] Medical science has brought us to an ultimate boundary beyond which a civilized and God-fearing society committed to the sanctity of marriage and the structure of the human family, may not go.'[36]

A number of church bodies, from the Roman Catholic Church to the Free Presbyterian Church of Scotland, took an uncompromisingly negative view of the practice of IVF, some even recommending that the practice be made a criminal offence. Basic to such positions was the stance that, since the embryo is inviolable from the time of fertilization, no embryos should be destroyed. Since this had happened in the development of IVF,

---

34 For a critical review of O'Donovan's book by another theologian see G. R. Dunstan, 'Review of Oliver O'Donovan. Begotten or Made?', *Religious Studies* 21/3 (1985), 415–16.

35 Oliver O'Donovan, *Begotten or Made?* (New York: Oxford University Press, 1984), 65, 85.

36 Thomas F. Torrance, *Test-Tube Babies: Morals – Science – and the Law* (Edinburgh: Scottish Academic Press, 1984), 1.

and was ongoing as a consequence of the production of surplus embryos in IVF programs, there could be no room for IVF within a Christian ethical framework. IVF was also regarded by some as an act of mastery in which human individuals are treated as little more than the subjects of technological domination, leading to the claim that the artificial production of children is dehumanizing. Some even expressed concern that an excessive use of IVF would lead to the bypassing of normal sexual intercourse as a means of producing children.[37]

The Pontifical Academy for Life, meeting in 2004, twenty-six years after the birth of the first IVF baby reinforced the official Roman Catholic position that the ARTs 'constitute an unworthy method for the coming forth of a new life, whose beginning depends [...] in large measure on the technical action of third parties outside the couple and takes place in a context totally separated from conjugal love'.[38] For the Academy, a child obtained through the use of the ARTs is on the same level as a product whose value depends upon its good quality. Surplus embryos in IVF programs are viewed as suffering an absurd fate since those who ordered them have rejected them.

While opposition to IVF and the ARTs in general represented one strand of Christian response, it was not the only one. For instance, the Board for Social Responsibility of the General Synod of the Church of England in a 1985 report supported all the recommendations of the Warnock Report, except embryo donation and the production of embryos for research purposes.[39] It acknowledged that there was division of opinion among members of the Board, differences that stemmed from divergent

---

37 B. McCarthy, *Fertility & Faith: The Ethics of Human Fertilization* (Leicester: Inter-Varsity Press, 1997), 198–203.

38 Pontifical Academy for Life, 'Final Communiqué on "the Dignity of Human Procreation and Reproductive Technologies. Anthropological and Ethical Aspects"', (February 2004) <http://www.vatican.va/roman_curia/pontifical_academies/acdlife/documents/rc_pont-acd_life_doc_20040316_x-gen-assembly-final_en.html> accessed 3 April 2013.

39 Board for Social Responsibility (Working Party on Human Fertilization and Embryology), *Personal Origins* (London: CIO Publishing, 1985).

theological viewpoints and ethical stances. What is of particular interest is the following comment:

> Some of our reasoning may seem to some people to be radical in light of past thought and practice before the new technologies had become possible. We wish to affirm however the large area of common ground among us all on which our reasoning is based. We take our stand on convictions about the nature of human life in the image of God and our duty to respect it. We are united in our conviction about the importance of marriage and the family and our need to uphold and support them. We are united in our commitment to ensure that informed and sympathetic pastoral help is offered to those who suffer from childlessness or infertility.[40]

Underlying this stance was a willingness to accept that, during its first fourteen days of development, the human embryo is not entitled to the same respect and protection as an embryo implanted in the uterus. This also allowed the Board to approve the use of surplus embryos in IVF programs for research purposes.

This stance was a welcome respite from the anti-reproductive technology furore evidenced by other theologians in the UK. However, as one reflects on some of the now historic statements from thirty years into the future, one has to ask whether one is being fair in raising them again. Any of us can jump to precipitate conclusions that we later regret. For this reason the next section outlines more recent responses. Before turning to these, it should be noted that O'Donovan's position in 1984 reflected his stringent criticism of what he viewed as the inroads of a technologically-inspired worldview, that gave the impression of leaving little or no room for any individual technological procedures. He severely criticized what he described as 'the sad story of Christian medicine in the last quarter century', and so for him IVF had to be rejected along with all the other reproductive technologies. If not, one would simply be driven to accept more and more of the technologies under the constraints of social pressure.[41]

---

40   Board for Social Responsibility (Working Party on Human Fertilization and Embryology), *Personal Origins*.

41   Oliver O'Donovan, 'A Neutered Morality', *Third Way* September (1984), 27.

Responses to the ARTs post-2000

It is now almost thirty years further on since *Manufacturing Humans* was published, and so it is important to ask: what has changed? The science has moved on in dramatic fashion. IVF is commonplace, PGD has entered clinical practice, human embryonic stem cells continue to be discussed endlessly even if their potential remains to be realized, therapeutic cloning has entered the research arena, and inter-species embryos are on the horizon. To what extent, if any, have Christian perspectives been influenced by the scientific developments? In other words, is the theological debate today markedly different from that of 1987?

My conclusion is that little has changed. The contemporary theological scene is remarkably similar to the theological scene of 1987. Those prepared to accommodate to the challenges presented by IVF are prepared to accommodate to the challenges of PGD, embryonic stem cells and therapeutic cloning. Those who saw IVF as entering illicit divine territory are appalled at subsequent developments that only serve to demonstrate that their initial repugnance was justified.

Roman Catholicism

This position has been most clearly expressed by the Congregation for the Doctrine of the Faith, with its 2008 publication of the instruction *Dignitas Personae on Certain Bioethical Questions*.[42] This was intended to bring up-to-date the earlier instruction *Respect for Human Life (Donum Vitae)*, published in 1987, in which the Vatican had set out its objections to IVF primarily on the grounds of its use of artificial means to achieve conception without sexual intercourse.[43] Interestingly, the more recent *Dignitas Personae* does not challenge the artificiality of the process, but

---

42 Congregation for the Doctrine of the Faith, 'Dignitas Personae'.
43 Congregation for the Doctrine of the Faith, 'Donum Vitae'.

seeks to protect human life on the grounds of its being personal from conception onwards. In arguing like this it aims to defend the dignity of the human embryo.

Its objection to the use of ICSI indicates a resistance to this technology because it 'establishes the domination of technology over the origin and destiny of the human person' and therefore is 'contrary to the dignity and equality that must be common to parents and children'.[44]

Other reproductive techniques are rejected on a variety of grounds. For instance,

- the freezing of embryos is rejected on pragmatic grounds – the process may harm them,
- the freezing of oocytes is rejected because it indirectly permits subsequent artificial reproduction,
- PGD is equated with eugenics and therefore completely ruled out,
- any destruction of embryos in IVF, let alone in research, is rejected because it represents injustice against embryos, and
- the donation of embryos is disallowed since this leads to illicit heterologous family relationships.

That is one side of the story, and yet it is only part of the story. Some serious Roman Catholic thinkers reject the version promulgated by the Vatican. A startling counterpoint is provided by the writings of two Roman Catholic bioethicists, Thomas A. Shannon and James J. Walter in their 2003 book, *The New Genetic Medicine*.[45] Their writings are exemplary for their willingness to wrestle with new scientific findings and directions, and with the possible implications of these for traditional formulae, including magisterial teachings of the Roman Catholic Church.

The authors contend that an individual is not present until about two-three weeks after the beginning of fertilization. They maintain that,

44   Congregation for the Doctrine of the Faith, 'Dignitas Personae'.
45   T. A. Shannon and J. J. Walter, *The New Genetic Medicine: Theological and Ethical Reflections* (Lanham, MD: Rowman & Littlefield, 2003).

although the preimplantation embryo possesses the genetic information necessary for development, this genetic uniqueness indicates the embryo's human nature, rather than an individuality that could claim moral privilege. Along these lines they argue that it is not until about three weeks after fertilization that the embryo can be deemed a physical individual, and it is not until individualization has occurred that the embryo can be considered a person. Hence, they assert that because the early embryo is not an individual person from fertilization, it cannot claim absolute protection. However, they also stress that the early embryo is valuable. This value is due to the embryo's genetic uniqueness and the ensuing human nature that is common to us all. Its value is thus independent of characteristics such as intelligence or relationality.

These arguments lead them to conclude that the preimplantation embryo has what they designate 'premoral' value. They argue that this premoral value should be weighed against other moral goods, such as any benefits arising from embryo research. This leaves open the possibility of conducting research on embryos, including embryonic stem cell research and therapeutic cloning, before individualization occurs around three weeks after fertilization. However, they do not give *carte blanche* to all embryo manipulation, but take care to specify limits and constraints, based upon their underlying theological vistas.

Shannon and Walter make a resolute effort to reinterpret traditional theological standpoints in the light of contemporary scientific thinking. They do not consider that this constitutes a divergence from tradition, which they argue was based in part on scientific theories that are now obsolete. Thus, their approach does not weaken the religious tradition with which they are imbued, but reaffirms it in contemporary terms.

Lest it is concluded that Shannon and Walter are errant exceptions within Roman Catholicism, it should be pointed out that various Roman Catholic scholars have roundly criticized the traditionalism represented by *Dignitas Personae*. For instance, Celia Deane-Drummond is concerned with the gap that has opened up between official pronouncements and the pastoral care needed for those faced with infertility.[46] She also critiques its emphasis on the alleged problems for human flourishing created by the donation of gametes or embryos, in the absence of any empirical assessment of the extent of these problems. She advocates an alternative approach to the reproductive technologies, based on a recovery of prudence that takes account of feminist concerns, and especially an ethic of feminist care. She wishes to see stress laid on the importance of caring for those with disabilities, including genetic disabilities.

Ann Marie Mealey's criticism of *Dignitas Personae* focuses on its reliance on what she regards as an outdated physicalist version of the natural law, and on the document's excessive concerns with a 'eugenic mentality'.[47] In contrast she would like to have seen it reflect upon what values we ought to be prioritizing in the face of current discussions about procreative procedures and stem cell research. For Mealey the document merely succeeds in reiterating and reaffirming older teachings rather than attempting to lay the groundwork for ways of responding to developments in genetics and for means of protecting the 'common good'.

While these three Roman Catholic contributions differ markedly in their respective features, they all take seriously contemporary scientific and clinical developments and they expect the religious contours they bring to the debates to address these. Not one of these writers wishes to dilute the core credentials of the Christian faith, but they do wish to make them relevant. The contrast between these contributions and the official Vatican stance is stark and points to a message those of other religious persuasions should take on board.

The crux of their message is that all human beings should be valued and treated as subjects worthy of human dignity. This applies to all involved in

---

46   Celia Deane-Drummond, 'Bodies in Glass: A Virtue Approach to Ethical Quandaries in a Cyborg Age through a Recovery of Practical Wisdom', in D. Gareth Jones and R. John Elford, eds, *A Glass Darkly: Medicine and Theology in Further Dialogue* (Bern: Peter Lang, 2010), 61–79.

47   Ann Marie Mealey, '*Dignitas Personae*', in D. Gareth Jones and R. John Elford, eds, *A Glass Darkly: Medicine and Theology in Further Dialogue* (Bern: Peter Lang, 2010), 111–29.

reproductive procedures, from embryos to women, and the networks binding all participants together. In arriving at this message they take serious note of ethical approaches not limited to the alleged status of the embryo, and incorporating additional notions such as prudence. They also reject conclusions formulated almost entirely on concern for the very earliest stages of embryonic development.

Protestantism

The divisions encountered within Roman Catholicism are, not surprisingly, evident within Protestantism. What might be surprising is that the intense suspicion of the ARTs, and even hostility towards them, encountered in the 1980s is taken in some circles today as representing the Christian view. This is especially prominent within evangelical circles.

A particularly detailed, but relatively little known, exposition of this hostility comes from Edwin Hui, Professor of Biomedical Ethics and Christianity and Chinese Culture at Regent College, Canada. He works out his position very thoroughly in his 2002 book *At the Beginning of Life: Dilemmas in Theological Ethics*.[48] The writer is located at an evangelical theological college and a leading American evangelical publisher produced the book.

Hui's basic theological postulate is that the human soul is present at conception, and that potentialities and capacities are given from the moment of God's creative act. Hence one is a person from fertilization onwards, and personhood is conveyed irrespective of one's inability to respond to God on account of one's iniquity, developmental immaturity or disability. A major outworking of this framework is Hui's opposition to any technological inroads into the reproductive process, including artificial contraceptives, the whole gamut of the ARTs, embryo manipulation and surrogacy. In his view use of the ARTs forces God to accept the child when

---

48 E. C. Hui, *At the Beginning of Life: Dilemmas in Theological Bioethics* (Downers Grove, IL: Inter Varsity Press, 2002).

he has not given that gift of life. He vigorously opposes use of embryonic stem cells, since from his perspective this reflects a serious disregard for the value of embryonic lives. His theological stance renders unacceptable any study of human embryos.

Hui though is not alone in the position he adopts, since there are many evangelical exponents of similar prohibitionist perspectives, even if none of them is elaborated with the care or at the length of Hui. They all stem from an embryo protection framework, and consequently have major repercussions for the extent to which they are or are not prepared to embrace the ARTs. This can be illustrated by various responses within the UK.

Stephen Bellamy, an evangelical Anglican vicar and bioethicist, undertook a detailed survey of the manner in which evangelical Christian groups responded to the UK debate on the 2008 Human Fertilisation and Embryology Bill.[49] Even a cursory scan of the more populist evangelical literature points to an almost unanimous view opposing PGD, tissue typing to obtain a saviour sibling, and the formation of cybrid and hybrid embryos. The overwhelming impression given by the information and briefing packs used to inform church leaders and members was that there was only one view of the embryo's status open to evangelicals who took the Bible and Christian faith seriously, and this was its complete protection from fertilization onwards.

An absolutist view of the status of the *in vitro* embryo undergirded this medley of stances held by groups such as the Christian Medical Fellowship (CMF), Christian Action Research and Education (CARE) and the Christian Institute (CI) in the UK. While there is not such unanimity on IVF itself, one gains the impression that a cautionary approach is by far the preferable one. It is noticeable that alternative viewpoints on the embryo and hence on the ARTs held by evangelical writers with a scientific background are ignored by Christian groups such as these in the public arena.[50]

49   Bellamy, 'Evangelicals and Embryology'.
50   Examples of such writers include Robert James Berry, *God and the Biologist* (Leicester: Apollos, 1996); Bryant and Searle, *Life in Our Hands*; James C. Peterson, *Changing Human Nature: Ecology, Ethics, Genes, and God* (Grand Rapids, MI: Wm. B. Eerdmans, 2010).

The problem from Bellamy's perspective is that, as a minister with pastoral responsibilities, the apparent certainty and rigidity of the absolutist views on the embryo creates unnecessary heartache and pain for evangelical Christians (and others). This applies not just to those contemplating PGD or tissue typing but also to the much larger group of those faced with the challenges of infertility. He writes: 'When these Christians ask at their church whether they are allowed to try using IVF in order to have a child, they are likely to be told that IVF is not open to Christians without being disobedient to God'.[51]

This concern is accentuated by the practice of referring to a prohibitionist view as representing *the* Christian worldview. According to one evangelical pastor all alternatives to an absolutist and prohibitionist view are to be overthrown: 'We [...] must be unashamed and unafraid for [...] unbiased medical technology is on our side. Time is on our side. The Bible is on our side. God is on our side. And if God be for us, who can be against us?'[52] The assurance of this stance is alarming, both regarding interpretation of the theology (see chapter 2) and even more so of the science. The fact that there are other views conscientiously held by other Christians is dismissed. An assumption that the Christian position is a prohibitionist one provides the context within which IVF is discussed and frequently dismissed in many spheres. For example, Dr. R. Albert Mohler, Jr., president of The Southern Baptist Theological Seminary, has called the destruction of embryos in IVF a tragedy. According to him, all involved in this technology are responsible for this vast human tragedy, intended or not. And, according to him, far too many evangelicals turn a blind eye to it and to what he describes as the harsh and grotesque reality that human life is destroyed as a result of using this technology.[53] Mohler concludes:

51   Bellamy, 'Evangelicals and Embryology', 188.
52   R. T. Kendall, 'The Sanctity of Life: A Biblical View', quoted in Bellamy, 'Evangelicals and Embryology'.
53   R. Albert Mohler, 'Christian Morality and Test Tube Babies, Part One', (29 September 2005) <http://www.albertmohler.com/2005/09/29/christian-morality-and-test-tube-babies-part-one-2/> accessed 17 June 2013; R. Albert Mohler, 'Christian Morality and Test Tube Babies, Part Two', (12 May 2006) <http://www.albertmohler.

IVF technologies destroy even as they claim to create, and the termination and disposal of human embryos is a reminder that the gruesome reality of the Third Reich
is never far from us [...] We must oppose the denial of human dignity to the unborn
and often forgotten embryos. We must oppose the use of these technologies by those
who would subvert the family, the marital covenant, and the Creator's gift of sexual
union and procreation.[54]

While there are elements that need to be taken seriously in these viewpoints,
the thrust is to condemn IVF and hence place it out of bounds for faithful
Christians. This is not to deny that the existence of surplus embryos creates problems for some parents as they lament the fate of their 'children'
or their chance of having another child. The manner in which embryos in
a freezer are viewed existentially as their babies points to an intuition that
should be regarded with a substantial degree of seriousness. Such couples
require support and understanding equivalent to the support they crave
when faced with infertility, with and without technical intervention.

An outspoken advocate for an alternative approach is that of
Presbyterian theologian, Ted Peters. He has come out most forcefully
with his 2008 book written in conjunction with two other theological
ethicists, *Sacred Cells? Why Christians Should Support Stem Cell Research*.[55]
As these writers look out on the scientific landscape, with its concentration
on stem cells, they identify three contending frameworks: embryo protection, human protection and future wholeness. In his many writings, Peters
is curious as to why so many theologians, when confronted by the value of
the human embryo, are drawn to the past, since this confines the debate to
what he considers is a confused account of genetic origin.[56] He contends

---

com/2006/05/12/christian-morality-and-test-tube-babies-part-two/> accessed 17
June 2013.

54   Mohler, 'Christian Morality and Test Tube Babies, Part Two'.
55   Ted Peters, Karen Lebacqz, and Gaymon Bennett, *Sacred Cells? Why Christians
      Should Support Stem Cell Research* (Lanham, MD: Rowman & Littlefield Publishers,
      2008).
56   Peters, *Playing God*; Ted Peters, *Science, Theology and Ethics* (Aldershot: Ashgate,
      2003); Ted Peters, *The Stem Cell Debate* (Minneapolis, MN: Fortress Press, 2007);
      Peters, Lebacqz, and Bennett, *Sacred Cells?*

that this is not required by Christian theology, since it omits reference to God's eschatological call to become who we are destined to be. This is closely allied with gifts given us by God, namely, our creativity as human beings, the glimpse we have been given of God's promised future, and our ability to make decisions for the good.

For Peters, while God confers human dignity, it is the duty of human beings to confer it on others. Hence when we confer dignity on someone we love, we treat that person as having intrinsic value. We treat the beloved as an end and not as a means to some further end. In these terms dignity is relational in character, in that it is the fruit of an ongoing loving relationship, expressed so clearly in the developing relationship between a mother and her newborn. This is where Peters' future orientation enters the picture, since he regards dignity as being derived more from destiny than from origin. The conferring of dignity on someone who does not yet experience or claim it is a gesture of hope. It is the future end product of God's saving activity rather than something imparted with the genetic code. This has ethical implications, since we are to impute dignity to those who may not already experience it, enabling them to claim it for themselves.

A second characteristic of Peters' position is that, since the spotlight is no longer directed exclusively onto the early embryo, the principle of beneficence can be included in ethical calculations. This allows him to examine those other groups who might benefit from a greater understanding of the embryo, emanating possibly from research on the embryo. Without this shift of the spotlight, non-maleficence towards the embryo forcefully trumps beneficence towards others within the human community. His aim is to rescue beneficence from the shadowy position it occupies in much theological thinking about reproductive issues.

What we have in Peters' position is a melding of divine action in conferring dignity, and human response in claiming dignity and ensuring that individuals are provided with an opportunity to blossom and flourish. In no way is dignity regarded as an automatic outworking of genetic characteristics. This follows from the future directedness of his position, together with the acknowledgement of human creativity and therefore of the centrality of human action. However, all this is viewed within a strong theological framework arising out of God's love for all, and an overarching eschatological hope based on God's promises of a coming kingdom of justice and fulfillment. Consequently, for Peters, the good of others in the community may on occasion trump the good of embryos.

A fascinating array of Protestant views on the embryo is contained in a 2003 edited volume, *God and the Embryo*, with four perspectives by Brent Waters, James Peterson, Ronald Cole-Turner, and Robert Song.[57] These range from reasonable doubt that embryos are ever to be subject to lethal research (Song), to permission under carefully regulated and limited conditions (Cole-Turner). Cole-Turner expresses the exasperation felt by many as he looks at the disagreements within the Christian community and sees no way out of this impasse over the embryo and what we owe it.[58]

Dominant themes

This overview of a range of Christian commentators should have brought out the main thrust still evident in ongoing debate on the reproductive technologies. While it is relatively easy to recognize these as the status of the embryo and the character of the reproductive technologies, there is far more interplay between the two than often acknowledged. Complicating matters further are views on the way in which contraception and abortion have transformed social and moral attitudes. If the permissiveness associated with these is found to be alarming it is but a short step to concluding that the same will happen with the reproductive technologies. As they become increasingly accepted, embryos will be increasingly used as commodities, family patterns will break down even further and children will be manufactured as products. Consequently, the use of embryos in

---

57   Waters and Cole-Turner, eds, *God and the Embryo*.
58   R. Cole-Turner, 'Principles and Politics: Beyond the Impasse over the Embryo', in
      B. Waters and R. Cole-Turner, eds, *God and the Embryo: Religious Voices on Stem
      Cells and Cloning* (Washington, DC: Georgetown University Press, 2003), 88–97.

technological ventures cannot be separated from radical changes within society. One does not have to be a Christian to come to this sort of conclusion, since commentators such as Leon Kass argue against the move from natural human procreation to manufacture. In discussing genetic manipulation, Kass writes:

> Make no mistake: the price to be paid for producing optimum or even only genetically sound babies will be the transfer of procreation from the home to the laboratory and its coincident transformation into manufacture [...] And let us not forget the powerful economic interests that will surely operate in this area; with their advent, the commodification of nascent human life will be unstoppable.[59]

For Kass what is important is the natural: human nature as we find it and not as humans would like it to be. Hence, his long-standing and implacable opposition to any form of IVF. Kass is not arguing from a Christian base, and yet his innate conservatism mirrors that of Christian writers who stress the centrality of the precautionary principle and the benefit-of-the-doubt approach. While the latter undoubtedly have a part to play in ethical debate, they must be seen for what they are – protectionist responses to permissive forces. They are seeking to protect both embryos and family structures in the face of liberalizing elements that may have little interest in protecting either of them.

What are we to make of these aims from a Christian perspective? I believe they are worthy aims, but do they lead to the conclusion that the reproductive technologies in every guise are to be rejected? By the same token one could equally well ask whether this also leads to a rejection of abortion under all circumstances and the rejection of all forms of (artificial) contraception? Before coming to such a conclusion one has to take into account other parties who would be affected by blanket rejection along these lines – those wishing to limit the size of their families as much as those longing for a child, and those confronted by horrendous illnesses in both mother and embryo/fetus. Neither would there be room for research

59   Leon R. Kass, 'Triumph or Tragedy – the Moral Meaning of Genetic Technology', *American Journal of Jurisprudence* 45 (2000), 1–16.

into early developmental patterns and embryonic growth, nor research into the causes of infertility. Not one of these possibilities is self-justifying, but all would be automatically excluded from consideration if the sphere of reproductive technology were to be closed down on ethical grounds (and theological grounds if Christians are advised against participating in any of them).

The context within which the reproductive technologies are approached extends well beyond embryos and fertility considerations, since it is part of the much broader endeavour that is modern medicine. This brings us back to O'Donovan's condemnation of the character of modern medicine, and while this has ramifications well beyond this book, it introduces an additional element, namely, that some concerns about the reproductive technologies stem from this wider perspective. However, it is one thing to criticize aspects of modern medicine and another to condemn the venture as a whole, especially if this condemnation has an allegedly Christian foundation. Unless one adopts the position that all modern medicine is to be rejected, and similarly the reproductive technologies, in favour one imagines of 'natural' therapies, there is no escape from the task of discerning what is to be accepted and what rejected. In other words, serious decisions have to be made, based on ethical and theological grounds (as outlined in chapter 2), between remedy (drug) A and remedy (drug) B, between aggressive and palliative treatment, between the need for treatment against refraining from it. These are case-based decisions, meeting individual needs and aspirations, when the path taken will generally reflect the value systems of the patients and their families. It is also worth pointing out that 'natural' therapies have to be assessed scientifically, clinically, ethically and theologically, by asking whether they benefit human beings.

The realms of natural therapies have their own built-in challenges, whether one is contemplating homeopathy, chiropractic, naturopathy, aromatherapy, Ayurveda or whatever.[60] Each has to be assessed on its own merits, especially if there is the possibility of being used apart from

60   Robina Coker, *Complementary and Alternative Medicine: Should Christians Be Involved?* (London: Christian Medical Fellowship, 2008).

conventional medical therapy. It is highly unlikely that any of them will emerge as problem-free. Do they bring healing and do they always serve the best interests of patients? Natural therapies can let people down, and if they are resorted to when more effective medical therapies are available, it is patients who will suffer.

In this chapter I have touched on a range of positions on the ARTs and associated procedures by a variety of Christian groups and church bodies. In their different ways all would justify their stance by reference to biblical teaching, one or other theological tradition, and/or Church pronouncements. All make some reference to science, although their acknowledged dependence upon scientific input varies immensely.

Positions purportedly having a biblical foundation often give the impression that they are derived from Scripture and are faithful to Scripture. Frequently however, they are just as dependent upon concepts that do not emanate from Scripture. Examples are the use of the precautionary principle and the related benefit-of-the-doubt argument; the genetic distinctness of the embryo, including its unique genotype; the dangers associated with the unknown risks of certain procedures; distortion of the population balance and threatening a child's identity in sex selection; and the lack of evidence demonstrating that embryo research has led to any cures. While these and arguments like them may have certain validity and should play a part in decision-making, they do not emanate from Scripture. Some are purely pragmatic and their rightness or wrongness will be ascertained empirically. Some are highly debatable and have long been the subject of ardent ethical debate. Some are essentially scientific comments, and have to be judged scientifically. Unfortunately, the injection of extra-biblical approaches such as these is frequently not acknowledged, and this is a severe hindrance to open debate within Christian circles.

Additional arguments often encountered in evangelical debate have a more ostensible theological rationale. These include claims such as: Christian ethics are based on God's revelation rather than human wisdom; Christian hope rather than medical intervention should be given pride of place; all human life is made in God's image and this is irrevocably implanted at fertilization; all human life is equally valuable especially that of exploited groups; children are a special gift from God, who opens the wombs of the infertile; and the value of an embryo is determined by God's grace and not by human decision-making

These theological statements have two characteristics: they are open to competing interpretations, and they fail to provide specific options for bioethical practice. While I have considerable sympathy with some of them, their generality is both their strength and their weakness. They are important in directing our attention to certain higher order theological principles, but in practice their thrust comes from their dependence upon the first set of arguments with their scientific or empirical bent or both. While they give the impression that they are dependent upon Scripture and/or a well-defined theological tradition, this assumption does not stand up well to critical scrutiny.

Since in my view it is not possible for any theological position to be completely independent of scientific input, it has to be determined what role science plays in the process of arriving at a theological position. It also follows that different theological positions will reflect differences in scientific understanding and interpretation just as much as they reflect differences in theological worldviews. If this is indeed the case, changes in certain aspects of a theological perspective would be expected to occur following major changes in some scientific notions. This may be unacceptable to those who hold to the view that all Christian perspectives are unchanging. In my view though, it is the starting point for productive discussion.

A final concern is that the reproductive technologies are largely undertaken by a commercial fertility industry, which is profit-driven and expansionary in outlook and flourishes in largely liberal secular societies. The result is that an increasing array of procedures is introduced and marketed and that these may enter clinical practice with inadequate research on their efficacy and safety.[61] These are legitimate concerns that have to be addressed by government agencies as well as professional bodies. There should be no

61   Rachel Brown and Joyce Harper, 'The Clinical Benefit and Safety of Current and Future Assisted Reproductive Technology', *Reproductive BioMedicine Online* 25/2 (2012), 108–17.

leniency in demanding the highest of ethical standards as well as stringent and well-formulated regulations.

There is also the reality that they are being used in situations where they have the potential for fragmenting families and for crossing generational boundaries. This is because we live in societies in which instant gratification and serving one's own interests outweigh living for others, serving others, accepting the givenness and giftedness of our existence, and being deeply concerned for the poor and neglected.

This is the problem facing all who attempt to bring their theological perspectives to bear on the ever-burgeoning reproductive technologies, since their *development* is scientifically driven whereas their *application* is community driven. The former is largely dependent upon the ethos of science, whereas the latter depends upon the worldview and hence interpretive framework of the community. However, these pressures also exist in other areas of health care, as well as beyond.

Using the reproductive technologies to create highly unusual and unstable family structures stands in stark contrast to central Christian imperatives. It is at this point that theology should be making its presence felt, rather than in constantly reworking arguments as to when embryos acquire absolute moral value. A Christian commitment should be directed at arguing for ways in which the technologies should be applied rather than in whether the technologies should or should not exist.

In this context it is useful to bear in mind the words of the preacher in Ecclesiastes 11:5, 'Just as you do not know how the breath comes to the bones in the mother's womb, so you do not know the work of God, who makes everything'. We live in the midst of profound uncertainties, and on so many occasions our insights into the work of God are deficient. Balancing the welfare and aspirations of competing groups is far from easy, requiring discernment, wisdom and a willingness to go on learning in an ever-changing world. The perplexity of our beginnings is set to increase, rather than decrease.

CHAPTER 4

Genes and self-image

The 'gene for X' story

Although this chapter is dealing with genetic issues, it will soon emerge that it inevitably touches on the deeper questions of how we view ourselves, what control we have over our actions, thoughts, hopes and fears, and where we are heading. While questions like this do not simply boil down to genetics, they are brought to a head by the prospects opened up by genetics, the major one of which is that of genetic determinism.

The media is saturated with reports of discovering a gene for such things as alcoholism,[1] homosexuality,[2] leadership,[3] obesity,[4] IQ,[5]

---

1   Subhash C. Pandey et al., 'Partial Deletion of the cAMP Response Element-Binding Protein Gene Promotes Alcohol-Drinking Behaviors', *The Journal of Neuroscience* 24/21 (2004), 5022–30.

2   Dean H. Hamer et al., 'A Linkage between DNA Markers on the X Chromosome and Male Sexual Orientation', *Science* 261/5119 (1993), 321–7; Stella Hu et al., 'Linkage between Sexual Orientation and Chromosome Xq28 in Males but Not in Females', *Nature Genetics* 11/3 (1995), 248–56; George Rice et al., 'Male Homosexuality: Absence of Linkage to Microsatellite Markers at Xq28', *Science* 284/5414 (1999), 665–7.

3   Jan-Emmanuel De Neve et al., 'Born to Lead? A Twin Design and Genetic Association Study of Leadership Role Occupancy', *The Leadership Quarterly* 24/1 (2013), 45–60.

4   K. Clément et al., 'A Mutation in the Human Leptin Receptor Gene Causes Obesity and Pituitary Dysfunction', *Nature* 392/6674 (1998), 398–401.

5   M. J. Chorney et al., 'A Quantitative Trait Locus Associated with Cognitive Ability in Children', *Psychological Science* 9/3 (1998), 159–66; Paul J. Fisher et al., 'DNA Pooling Identifies QTLs on Chromosome 4 for General Cognitive Ability in Children', *Human Molecular Genetics* 8/5 (1999), 915–22; Robert Plomin et al., 'A Genome-Wide

aggression,[6] religiosity,[7] moral awareness[8] etc. There even appear to be smart mice, on account of their genes.[9] No matter what feature one thinks of, there appears to be a gene for it, or there are claims that this is the case. All such claims are based on the underlying assumption that there is a direct correlation between genes and disease, genes and behaviour, and even genes and belief.

From here it is but a short step to the ability to choose genes for our children, rendering them intelligent, bright, beautiful and possibly even virtuous. On the other hand, if one continues to track along this path, one can speculate that a despot would have the power to produce aggressive and manipulative beings fit for combat and destruction. Even putting these unpleasant scenarios to one side, the hope appears to be that we could ensure that genetically modified individuals turn out to be compliant to our wishes, becoming entrepreneurs, scientists or accountants, or excelling in chess, football or ballet. Perhaps we would have within our grasp the power to make them religious, or in precisely the same way irreligious. For every religious gene one can imagine an atheist gene. The end results are of little consequence for this debate; what is crucial is the alleged power of genes and genetic selection.

As I sit at a computer writing these words, I am forced to reflect that I may only be doing this because I have a gene for writing, perhaps even a gene for writing the sort of material I write. After all, I am not a novelist,

Scan of 1842 DNA Markers for Allelic Associations with General Cognitive Ability: A Five-Stage Design Using DNA Pooling and Extreme Selected Groups', *Behavior Genetics* 31/6 (2001), 497–509.

6   Han G. Brunner et al., 'Abnormal Behavior Associated with a Point Mutation in the Structural Gene for Monoamine Oxidase A', *Science* 262 (1993), 578–80.

7   Dean H. Hamer, *The God Gene: How Faith Is Hardwired into Our Brains* (New York: Doubleday, 2004).

8   Abigail A. Marsh et al., 'Serotonin Transporter Genotype (5-HTTLPR) Predicts Utilitarian Moral Judgments', *PloS One* 6/10 (2011), e25148.

9   S. E. Lee et al., 'RGS14 Is a Natural Suppressor of Both Synaptic Plasticity in CA2 Neurons and Hippocampal-Based Learning and Memory', *Proceedings of the National Academy of Sciences* 107/39 (2010), 16994–8; Ya-Ping Tang et al., 'Genetic Enhancement of Learning and Memory in Mice', *Nature* 401/6748 (1999), 63–9.

neither am I a philosopher, and so it may be that I lack those particular genes. A little thought should make evident the absurdity of thinking in this way. The number of genes required to control every minor feature of our lives would soon outstrip the 20,000 genes in the human genome.

This way of thinking reflects the cult of the gene and its supposed explanatory powers. A gene will be found for every human feature, just as mice can be explained in purely genetic terms. It takes little imagination to conclude that we are our genes, but if this is the case, it appears to leave no room for anything else. We soon become 'nothing but' our genes, with every facet of what we are enshrined in our genetic make-up. In theological terms this would equate with the most hard-wired predestination one can imagine. While these claims are more at home in the popular media than in serious debate, they stand or fall in scientific terms. They are either essentially correct or they are misleading. This is of far more than theoretical interest, since if the claims made for genes are overhyped or even plain wrong, all other subsidiary claims are thrown in doubt.

What one is dealing with here is the 'gene myth',[10] which in the words of Peters refers to 'a thought form or conceptual set, a cultural frame through which we interpret the accelerating growth in scientific knowledge about DNA.[11] Its key undergirding assumption is genetic determinism. The compelling facet of the gene myth is its simplicity; it is akin to the one explanation of everything in the physical sciences. In one fell swoop it answers questions about who I am and why I am as I am, with clarity and assurance. According to the gene myth, one could argue, for example, that I am as I am because I was born with gene ltx45 (to make up a gene), since it is this gene that makes me assertive and precise; my leadership capabilities all stem from this gene, as does my ambition. Were I to lack this gene, I may be more creative and artistic, with little interest in having others follow my lead; I may also be more empathic. When expressed in

10 Ruth Hubbard and Elijah Wald, *Exploding the Gene Myth: How Genetic Information Is Produced and Manipulated by Scientists, Physicians, Employers, Insurance Companies, Educators, and Law Enforcers* (Boston, MA: Beacon Press, 1993).
11 Peters, *Playing God?*, 7.

these simple causal terms, no account needs be taken of environmental or
social factors, how I was brought up, what educational opportunities I have
had, the social networks of which I am part, nor the resources I have at
my disposal. Neither is there any need to take note of my own aspirations
and value systems, on the ground one imagines that these are all genetically
determined. But are they?

While it is tempting to indulge in these possibilities (or fantasies) as
theoretical exercises, the gene myth constitutes the backdrop for thinking
about genetic choices that are much closer to the clinic. A geneticist in a
paediatric clinic can carry out her role without having any sympathy for
the gene myth, being well aware that absolute certainty of all the causative
factors in a particular case is rarely possible without knowledge of a host of
environmental influences and even other unknowns. Alternatively, those
promulgating the importance of genes in making us what we are may have
the most rudimentary understanding of genetics and the genetic causes
of disease. This gulf is an important one when considering how to relate
grandiose genetic claims to actual decision-making about treatment options
for children with genetic disorders. It is important to avoid the situation
whereby opposition to grandiose claims translates into opposition to the
utilization of genetic knowledge in the clinic. This would mark a transition
of momentous proportions, from the level of theoretical debate to practical medical genetics. While philosophers and theologians are at home at
the first level, geneticists and health care practitioners are at home at the
second level. Caution is required to ensure that the idealistic gene myth
story does not intrude into the clinic and consulting room, where it has
no explanatory power and no place.

Personalized medicine

The decoding of the human genome in 2003 brought with it the prospect
of opening up unprecedented understanding of the functions and interactions of the entire genome, this in turn leading to a revolution in health

care.[12] This was to be the new world of personalized or genomic medicine. The great strides anticipated at the time lay in the treatment of multifactorial disorders, that is, breast and colorectal cancers, and Alzheimer's and Parkinson's diseases. Additionally, it was hoped that this new approach would uncover the mechanisms of complex diseases, including asthma, hypertension and diabetes.[13] The prospect of one's family doctor having at their disposal a vastly enhanced armamentarium of genetic diagnostic tools stemmed from a deterministic view of genetics, centered on the ability of geneticists and their clinical colleagues to translate the promise of genetic science into clinical reality.

Personalized medicine has long been seen as the therapeutic face offered by the human genome project, presenting as it does enticing therapeutic vistas. In theory, genetic knowledge of this order could enhance people's understanding of themselves and their world. For instance, instead of having to think vaguely about, say, cholesterol levels, which may or may not have the significance attributed to them for particular individuals, people would theoretically have a far more precise means of knowing whether these levels should be taken seriously in individual cases. Whether or not people could cope with such detailed information is another matter, since the medicalization of life may become overbearing.

However, even in a world characterized by this level of genetic foreknowledge, there would still be an intimate connection between people's genes and the numerous environmental factors to have influenced genetic expression since the first few days of embryonic existence. A strong predisposition to develop stomach cancer is affected by dietary, neuroendocrine, external environmental and attitudinal factors.[14] It is a person, and not a

12 A. E. Guttmacher and F. S. Collins, 'Genomic Medicine – a Primer', *The New England Journal of Medicine* 347/19 (2002), 1512–20.

13 Michael King, Maja Whitaker, and D. Gareth Jones, 'Speculative Ethics: Valid Enterprise or Tragic Cul-De-Sac?', in Abraham Rudnick, ed. *Bioethics in the 21st Century* (InTech, 2011), 139–58.

14 K. E. McColl, H. Watabe, and M. H. Derakhshan, 'Sporadic Gastric Cancer: A Complex Interaction of Genetic and Environmental Risk Factors', *American Journal of Gastroenterology* 102/9 (2007), 1893–5.

set of genes, who develops stomach cancer. In other words, even in some future world of genetic foreknowledge, the crucial context will still be that of people in their wholeness, and not genes in some aseptic, depersonalized cellular compartments.

As it has turned out, the gulf between genomics and personalized medicine is far greater than envisaged, resulting in overstated optimism and unrealistic timescales.[15] The harsh reality is that sequencing an individual's genome produces very large amounts of information, much of which is difficult to interpret. As a consequence, genomic medicine with its myriad possibilities has not as yet had a dramatic influence on day-to-day health care for individual patients.[16]

In spite of these considerations, the notion of the power of the gene is difficult to dispel, as forays continue to be made into investigating how it might be exploited. Consider the two following cases.

> Instead of the usual required summer-reading book, [a few years ago] incoming freshmen at the University of California, Berkeley, [were to] get something quite different: a cotton swab on which they can, if they choose, send in a DNA sample. The university said it would analyze the samples, from inside students' cheeks, for three genes that help regulate the ability to metabolize alcohol, lactose and folates. Those genes were chosen not because they indicate serious health risks but because students with certain genetic markers may be able to lead healthier lives by drinking less, avoiding dairy products or eating more leafy green vegetables. Berkeley's program, for the class of 2014 [was designed to be] the first mass genetic testing by a university [...] it was designed to help students learn about personalized medicine and identify their own vulnerabilities.[17]

---

15   F. Collins, 'Medical and Societal Consequences of the Human Genome Project', *New England Journal of Medicine* 341/1 (1999), 28–37; Guttmacher and Collins, 'Genomic Medicine – a Primer'; G. J. van Ommen, E. Bakker, and J. T. den Dunnen, 'The Human Genome Project and the Future of Diagnostics, Treatment, and Prevention', *Lancet* 354, Suppl. 1 (1999), S5–10.

16   F. Collins, 'Has the Revolution Arrived?', *Nature* 464/7289 (2010), 674–5; W. D. Hall, R. Mathews, and K. I. Morley, 'Being More Realistic About the Public Health Impact of Genomic Medicine', *PLoS Medicine* 7/10 (2010), e1000347.

17   Tamar Lewin, 'College Bound, DNA Swab in Hand', *New York Times* (18 May 2010).

This is an illustration of personalized medicine, the working assumption being that the individuals concerned, students in this case, will be able to make decisions based upon knowledge of their genetic status. The testing regime is non-invasive, and its aim is to prepare the students for the new genetics and the world of personalized medicine. In this instance, however, there were ethical concerns since it was to be carried out in the absence of counselling. And it was concerns of this nature that brought this idea to a premature halt. Additionally, how might students who tested negative have responded? Perhaps some of them would have concluded that they could drink more alcohol; after all they would not react negatively. While this is speculation, it demonstrates that genetic testing, even testing as apparently innocuous as in this case, does not exist in a social vacuum. It is people who are being tested and they are far more complex than the genes that help regulate the ability to metabolize alcohol, lactose and folates, or any other genes for that matter.

A somewhat different aspect of this same issue is raised by another news item.

> An NHS hospital in the UK has begun decoding all the genes of individual patients, 10 years after the first human genome sequence was published. London's Royal Brompton Hospital said the project would give doctors a better understanding of the inherited factors that help trigger heart disease. The research involves sequencing all 20,000 or so genes found in the human genome in 10,000 patients. It heralds more personalised treatments for diseases [...]
> They will also have a detailed MRI scan of the heart to show how it is functioning. The study has been made possible because of dramatic progress in the speed of DNA sequencing [...] Professor Dudley Pennell [...] professor of Cardiology [...] said: 'Ultimately our aim is for someone to come in and have a full scan and genetic analysis, leading to a personalised therapy which will treat their particular disease' [...] Patients have to undergo extensive genetic counselling before their results are revealed.[18]

---

18   Fergus Walsh, '10,000 NHS Patients "to Have Genes Mapped"', *BBC News* (21 June 2010) <http://www.bbc.co.uk/news/10367883> accessed 5 April 2013.

At specialized centres it is now possible to sequence an entire genome of an individual in thirteen hours or less (the length of time and the cost are decreasing rapidly each year).[19] The attraction is that as more patients are screened, the basis will be laid for understanding which genes are responsible for triggering disease. This is where the power of personalized medicine lies, since theoretically at least it will be possible to develop specific therapies for specific individuals or groups of individuals, based on knowledge of their genome.

While these news items move some distance from the 'gene for X' story, the similarity is only too clear. Their legitimacy is dependent upon the power of genes. There is a presumption that one's way of life will have limited influence on genes, and also that the knowledge of having a genetic predisposition to succumb to a serious disease will have little effect on one's life. However, the context for such decision making should be that we are people who make decisions on how to live, how we spend our time, what our priorities are, and what value systems we espouse. While we should take account of knowledge that our genetic constitution is predisposed towards illness D, the latter should not be viewed as an inevitable outcome regardless of how we live. Genes are important, and knowledge along these lines may serve an invaluable function in helping us to live appropriately, but emphasis upon them should not obscure the influence of other factors.

The deceptive use of genetic knowledge

Up to this juncture I have confined my attention to legitimate forays into personalized medicine. However, the power of the gene myth is such that the constraints typical of serious clinical trials and analyses no longer apply. If knowledge of my genetic make-up can give me information on how the future will turn out, and if I am offered such information for a few hundred dollars on the web, why not take it? This is the realm of direct-to-consumer (DTC) testing, with use of next generation sequencing technologies, the

---

19   Walsh, '10,000 NHS Patients "to Have Genes Mapped"'.

role of which is to test for susceptibility to hundreds of diseases and traits.[20] (Caulfield and McGuire 2012). Prominent DTC companies are 23andMe, Navigenics, and deCODEme.[21]

23andMe describes itself in the following way:

> 23andMe, Inc. is the leading personal genetics company dedicated to helping individuals understand their own genetic information through DNA analysis technologies and web-based interactive tools. The company's Personal Genome Service® enables individuals to gain deeper insights into their ancestry and inherited traits. The vision for 23andMe is to personalize healthcare by making and supporting meaningful discoveries through genetic research. 23andMe, Inc., was founded in 2006, and the company is advised by a group of renowned experts in the fields of human genetics, bioinformatics and computer science.

The increasing number of DTC companies (and the competition between them) and the increasing range of health-related services offered to the public has led to the appearance of a number of policy recommendations in different countries.[22] These have included in depth discussion of privacy issues associated with DTC companies, potential adverse effects on health care systems, the importance of providing accurate data, and the need for education in assisting the public interpret genetic risk information. The size of the market for these tests appears to be smaller than one might imagine, and yet it is not negligible especially as costs decrease.

DTC offerings emanate from, and derive their attraction from, serious laboratory and clinical science. But the provisos built into that science have been downgraded. Simplicity is the order of the day, since it tends to provide definite data and makes 'predictions that are medically unproven and so ambiguous that they do not provide meaningful information to

---

20   T. Caulfield and A. L. McGuire, 'Direct-to-Consumer Genetic Testing: Perceptions, Problems, and Policy Responses', *Annual Review of Medicine* 63 (2012), 23–33.

21   See http://www.23andMe.com, http://www.navigenics.com, http://www.decodeme. com.

22   See Caulfield and McGuire, 'Direct-to-Consumer Genetic Testing'.

consumers'.[23] For example, 'you have genes for M, D and Y, and they will give rise to conditions m, d and y'. Even if it is stated that the chances of this happening are 40 per cent, 25 per cent and 15 per cent, these figures are exceedingly tenuous and place the client in a predicament as to how to interpret them, especially in the absence of health professionals to assist. Some companies provide genetic counselling services to their consumers, but a recent survey found that only about 10 per cent of consumers reported using these services.[24]

This is a world eminently suited to the gene myth. It is so much easier to receive a print out with one's genetic data on it – hard data – rather than a recipe for good nutrition, a balanced lifestyle and adequate exercise. The latter requires work and discipline as does the genetic data if it requires modification of lifestyle, but the environmental aspect is very easily ignored.

Genes and the environment

An area of increasing interest over recent years is epigenetics that has been variously defined over the years. A typical definition is that it is 'the study of heritable changes in gene expression that are not due to changes in DNA sequence'.[25] More helpful from our angle is the notion that 'epigenetic states can become disrupted by environmental influences or during ageing', an observation of considerable significance in understanding the development of cancer and other diseases.[26] Central to this field is the idea that genes have a memory, and that the experiences of forebears in previous generations

23   Gregory Kutz, D., 'Nutrigenetic Testing: Tests Purchased from Four Web Sites Mislead Consumers', Government Accountability Office (27 July 2006) <http://www.gao.gov/products/GAO-06-977T> accessed 18 June 2013.
24   Caulfield and McGuire, 'Direct-to-Consumer Genetic Testing'.
25   Alex Eccleston et al., 'Epigenetics', *Nature* 447/7143 (2007), 395.
26   Eccleston et al., 'Epigenetics'.

may exert an influence over the present generation.[27] This far-reaching idea is based on various pieces of evidence, one of which is that monozygotic twins do not always show the same susceptibility to diseases, especially as they age.[28] Epigenetics, therefore, may provide an explanation for the impact of the environment on physiology and behaviour.[29] It explains how the same genetic sequence can be functionally altered by environmental influences during development and also probably ageing. According to one research institute, 'It explains why children with a comparable genetic heritage can grow up with their own individual phenotype which includes body composition and health risks'.[30] This provides a backdrop for looking at the relationship between genes and the environment.

Consider the following speculative scenario.[31]

> Josie is twenty years of age, and in our imagination let us sketch four ways in which she might have been brought up, remembering that the Josie of our story commenced life with a particular set of genetic characteristics. Josie 1 has been brought up in a loving home in an affluent neighbourhood; she has three siblings and has received an excellent education. Josie 2 has also been brought up in a loving home, but her family are poor and her father died when she was ten. They live on a barely satisfactory diet, there are three younger siblings, and Josie 2 had to leave school at sixteen in order to help financially with the family. Josie 3, like Josie 1, was brought up in affluence, but there is constant strife in the family, and her father is abusive towards her mother as well as her. Josie 3 is an only child and has had a good education. Josie 4 is also an only child, is encouraged constantly to perform well and is pushed educationally in order to be the doctor her father wished he had been. She was sent away to boarding school at the age of 10.

27   A. Bird, 'Perceptions of Epigenetics', *Nature* 447/7143 (2007), 396–8.
28   A. H. Wong, I. I. Gottesman, and A. Petronis, 'Phenotypic Differences in Genetically Identical Organisms: The Epigenetic Perspective', *Human Molecular Genetics* 14/ Suppl 1 (2005), R11–18.
29   A. P. Feinberg, 'Phenotypic Plasticity and the Epigenetics of Human Disease', *Nature* 447/7143 (2007), 433–40.
30   Liggins Institute, 'Developmental Epigenetics', <http://www.liggins.auckland.ac.nz/ uoa/home/about/research-themes/developmentalepigenetics> accessed 18 June 2013.
31   Jones, *Bioethics*.

The hypothetical Josies of these scenarios (versions 1–4) will differ markedly by the age of twenty, even though genetically they would have been identical.[32] Nevertheless, the contrasting environments in which they were reared and their different life experiences will ensure that the twenty-year-olds have only a limited amount in common. Of course they are recognizable as the same person, but they are far from identical. They may well have different aspirations and hopes, different interests and even different value systems. Their similarities will be due to their genetic moulding, and their differences to the formative influences under which they have grown up.

Speculative as this scenario is, and one that can never be tested, identical twins are relatively common and have been the subjects of myriad studies.[33] Their similarities and differences continue to be the subject of intense debate, but of one thing we can be certain, no matter how they have been reared and subsequently raised, they are not identical people, regardless of their identical genetic make-up.[34]

The interplay between genes and environment is intimate. All our genetic building blocks are far from unalterable, since the environment affects everything to which they give rise. The environment extends down to the cellular level.[35] In other words, genes should never be looked upon as isolated units, acting in predetermined fashion oblivious to a host of environmental influences. The implications of this are that genes are switched on and off indirectly as well as directly.[36] Advertently or inadvertently,

32 Jones, *Bioethics*.

33 Dorret Boomsma, Andreas Busjahn, and Leena Peltonen, 'Classical Twin Studies and Beyond', *Nature Reviews Genetics* 3/11 (2002), 872–82.

34 George Mandler, 'Apart from Genetics: What Makes Monozygotic Twins Similar?', *Journal of Mind and Behavior* 22/2 (2001), 147–59; H. H. Newman, F. N. Freeman, and K. J. Holzinger, 'Twins: A Study of Heredity and Environment', *Eugenics Review* 30/1 (1937), 61–2; R. Plomin and D. Daniels, 'Why Are Children in the Same Family So Different from One Another?', *International Journal of Epidemiology* 40/3 (2011), 563–82.

35 I. Lobo, 'Environmental Influences on Gene Expression', *Nature Education* 1 (2008), 1.

36 A. Ralston and K. Shaw, 'Environment Controls Gene Expression: Sex Determination and the Onset of Genetic Disorders', *Nature Education* 1 (2008), 1.

their functioning may be modified by the nature of the environment in which children grow up and function, and later in which adults function. None of us exists in a social vacuum; we (and our genes) are constantly being influenced by all manner of factors. Nothing is set in concrete, not even genes and their influence. Consider the following real life scenario.

> The criminal brain has always held a fascination for James Fallon. For nearly 20 years, the neuroscientist at the University of California-Irvine has studied the brains of psychopaths. He studies the biological basis for behaviour, and one of his specialties is to try to figure out how a killer's brain differs from that of most other people, the non-killers in the world.[37]
>
> A few years ago Fallon made a startling discovery. He knew he had a dubious family history, stretching back many generations. There was evidence that there had been a number of killers in his family over the generations. Fallon was well aware that the orbital cortex in the brain puts a brake on the brain region known as the amygdala, a crucial region implicated with the control of aggression. Unfortunately, in some people, there is a malfunction and imbalance, leading to excessive rage and violent behaviour.
>
> Fallon had previously persuaded ten of his close relatives to submit to a PET (positron emission tomography) brain scan and give a blood sample as part of a project to ascertain whether his family had a risk for developing Alzheimer's disease. In light of his violent family history, he examined the images and compared them with the brains of psychopaths looking at how active or otherwise was the orbital cortex in these family members. To his relief his wife's scan was normal, as were those of his mother, siblings and children; but his own was not. It appeared that his orbital cortex was relatively inactive, meaning that it exerted less control over the amygdala.
>
> In view of this alarming finding, he proceeded further and tested each family member's DNA for genes associated with violence, by examining the monoamine oxidase A (MAO-A) gene, the so-called 'warrior gene' that regulates a neurotransmitter, serotonin, in the brain.[38] With a certain version of this 'warrior gene', the brain fails to respond to the calming effects of serotonin. He discovered that everyone in his family who was tested for this gene had the low-aggression variant of it, except for one person: himself. Since this gene has been repeatedly found to be associated

37 Barbara Bradley Hagerty, 'A Neuroscientist Uncovers a Dark Secret', *National Public Radio* (29 June 2010) <http://www.npr.org/templates/story/story.php?story Id=127888976> accessed 5 April 2013.

38 A. Gibbons, 'American Association of Physical Anthropologists Meeting. Tracking the Evolutionary History of a "Warrior" Gene', *Science* 304/5672 (2004), 818.

with excessive aggression, the gene myth explanation would be that he was capable of excessive aggression, possibly resulting in death as with some of his ancestors. But in spite of the presence of this gene (the so-called gene for aggression) and the neural characteristics of the appropriate brain region, he did not fall into this category. The reason is now well known: he had had a happy and fulfilled childhood.[39]

In other words, brain patterns and genetic makeup are not enough to make anyone a psychopath. A third ingredient has to be present, an environmental one: abuse or violence in one's childhood. Take this away and the first two will not be sufficient to produce a seriously aggressive individual. Even in a situation like this one, where the genetic foundation is primed to produce aggression, the presence of certain genes in isolation of other factors does not predetermine the behaviour of the individual.

It is into this discussion that legal issues are raised. Consider the following based on an original article in *New Scientist*.

> In 2007 Abdelmalek Bayout admitted to stabbing and killing a man and received a sentence of nine years and two months. In 2009 an appeal court judge in Trieste, Italy, cut the sentence by a year after finding out he has gene variants linked to aggression. Nita Farahany, a legal scholar, is concerned because genes do not 'tell us why they did the thing they did and that's what criminal cases are ultimately interested in'. What's more the gene argument seems to cut both ways. According to Farahany, US courts are increasingly using genes in evidence for the prosecution. 'People don't recognize the double-edged potential of this evidence.'[40]

This is a clear example of a judge reducing a killer's sentence on account of his genetic mutations linked to violence. From his angle these were sufficiently involved to override full responsibility on Bayout's part. The message to emerge is that one's genes can partly absolve one of responsibility for a particular act. This downplays any role that social and environmental factors may play in behaviour, let alone freewill and personal responsibility

---

39   A. Caspi et al., 'Role of Genotype in the Cycle of Violence in Maltreated Children', *Science* 297/5582 (2002), 851–4.

40   Ewen Callaway, 'Murderer with "Aggression Genes" Gets Sentence Cut', *New Scientist* (3 November 2009) <http://www.newscientist.com/article/dn18098-murderer-with-aggression-genes-gets-sentence-cut.html> accessed 7 April 2013.

for decision-making. The disconcerting aspect is that this legal judgement was based on an unreliable interpretation of the science regarding the context within which the crime was committed.[41]

It may be that some people are genetically inclined to be more aggressive than others, but this is no more than part of the story. By itself this does not lead to the conclusion that as individuals they are committed to being aggressive. An inclination is simply that, it does not amount to an inevitable way of behaving. Unfortunately, the situation is aggravated when such individuals have been brought up in an early environment characterized by physical abuse. For instance, abused children with certain genetic polymorphisms are more likely to show anti social behaviour.[42] But even this combination of genetic and social factors does not inevitably lead to this type of behaviour, although it predisposes to it. The take home message is that the role models provided by parents and siblings are all crucial determinants of whether particular genes will or will not be expressed. And of course what applies to aggression applies just as much to all other personality characteristics.

Taking this further, we can also say that these genes affect the manner in which people's brains develop, and these in turn are fundamental to the personal characteristics and qualities individuals demonstrate.[43] Choices are made by people and not by their genes or even their brains. Human beings cannot hide behind their genetic make-up, and argue that their genes made them do X rather than Y, to hit someone rather than care for them. Unfortunately, individuals brought up in seriously abusive and dysfunctional situations are probably at a huge disadvantage in this regard, and should not be categorically condemned. Nevertheless, people are responsible for their behaviour, even recognizing its major genetic and

---

41 Emiliano Feresin, 'Lighter Sentence for Murderer with "Bad Genes"', *Nature* 10 (2009), 1038.

42 Julia Kim-Cohen et al., 'MAOA, Maltreatment, and Gene–Environment Interaction Predicting Children's Mental Health: New Evidence and a Meta-Analysis', *Molecular Psychiatry* 11/10 (2006), 903–13.

43 Jones, *Bioethics*.

environmental components and the intimate genetic-environmental interactions that contribute to what we are as individual human beings.

Genetic determinism

Lurking in the background of this discussion is the concept of determinism,
which is the view that things happen because they were predetermined to
happen. According to Lewontin and co-authors, biological determinism
suggests that 'human lives and actions are inevitable consequences of the
biochemical properties of the cells that make up the individual; and these
characteristics are in turn uniquely determined by the constituents of the
genes possessed by each individual'.[44] Hence, an action such as the decision
to get out of a chair can be ascribed to a series of complex events with a
genetic basis. It was this series of events that determined that this decision
would be taken within the context in which it occurred. While it may make
sense to reduce complex actions to their simplest constituents, genetic in
this case, the move to make determinist predictions about behaviours and
actions of the whole on the basis of these smaller categories is the move
from methodological reductionism to epistemological reductionism.[45]
Reducing complex actions to their simplest constituents constitutes the
power of the scientific method, but it is a major leap from here to the proposal that genetic reductionism bestows upon genes explanatory powers
as to how we behave as we do. Can this move be justified scientifically and
what are its implications?

44   R. C. Lewontin, Steven Rose, and Leon J. Kamin, *Not in Our Genes: Biology, Ideology
     and Human Nature* (New York, NY: Pantheon Books, 1984), 6.
45   Nancey Murphy and Warren S. Brown, *Did My Neurons Make Me Do It? Philosophical
     and Neurobiological Perspectives on Moral Responsibility and Free Will* (Oxford:
     Oxford University Press, 2007).

Genetic determinism has problems of its own. There are not enough genes in the human body to code for all the cells and processes in a one-to-one manner. For instance, there are millions of capillaries in the body and there is no way in which every bifurcation in every capillary can have a separate gene.[46] It is not possible for every physical trait and behavioural attribute to be coded for by its own gene.[47] A second issue is linked to stochastic gene expression, or chaos, whereby phenotypic variation occurs among genetically identical cells exposed to the same environment. Raser and O'Shea argue that 'even cells or organisms with the same genes, in the same environment, with the same history, display variations in form and behaviour that can be subtle or dramatic'.[48] In light of these observations one has to question whether it is feasible that our thoughts and actions can possibly be inevitable consequences of the biochemical properties of cells.[49]

Why then the role given to genetic determinism by so many writers? In part the path from genes to proteins to physical characteristics is relatively direct and compelling, so that when there is a malfunction in the genes that code for a particular protein, a discernible error can be detected phenotypically. For instance, if little melanin is produced, the individual will have blue or light eyes.[50] It is the genes that code for the protein the melanocyte needs in order to create melanin.[51] As a result, a mutation in the DNA could result in the absence of a protein necessary to create melanin. Consequently, gene mutations reduce the amount of pigment made.

This simple explanation helps explain why a form of genetic determinism has proved so useful in genetic testing. Prenatal testing of various

---

46 R. Sapolski, 'Chaos and Reductionism', Stanford University (19 May 2010) <http://www.youtube.com/watch?v=_njf8jwEGR0> accessed 18 June 2013.

47 E. Johnston, 'Nature Versus Nurture: Are We Missing a Third Option?', (Thesis, Master of Health Sciences, University of Otago, 2013).

48 J. M. Raser and E. K. O'Shea, 'Noise in Gene Expression: Origins, Consequences, and Control', *Science* 309/5743 (2005), 2010–13.

49 Lewontin, Rose, and Kamin, *Not in Our Genes*.

50 N. Lamb, 'The Genetics of Eye Color', *Community Outreach* (Fall 2009) <http://www.hudsonalpha.org/sites/default/files/pdf/genetics_of_eye_color.pdf> accessed 18 June 2013.

51 Johnston.

descriptions for conditions such as Down Syndrome, neural tube defects, spinal muscular atrophy and cystic fibrosis has proved powerful, since there is a one-to-one relationship between the genetic defect and the resulting condition.[52] Nevertheless, even in these cases, the tests may not be sufficiently powerful or accurate to indicate the severity of the conditions, let alone how the individual affected, or their family, will respond to living with it. Even in these highly focused areas, therefore, where the benefits of genetic determinism are evident, determinism has its shortcomings.

These comments should serve as warnings when confronted by the speculative possibilities regularly debated around screening for complex traits like height, beauty, intelligence and sexual orientation.[53] While such aspirations have little basis in serious science, those who advocate moving in these directions are relying heavily on genetic determinism, with its assumption that the modification of one or more genes will produce the desired result – an increase in height or intelligence or whatever. Quite apart from ethical considerations, this reliance upon the notion of genetic determinism is deeply problematic, since there is no assurance that the end result will be what is desired.

More generally, genetic determinism may be employed to suggest that everything we do is determined, a product of the genetic arrangement present at fertilization. Superficial and misleading as this view is scientifically, it sends out powerful ideological messages: gene therapy will usher in a new age of health and longevity. Alongside this is the view that with further unravelling of the human genome, more and more profound mysteries of human existence will be amenable to human understanding and exploitation. Humans will indeed control human destiny in a way only barely imaginable at present, as touched on in chapters 1 and 2.

52   S. M. Dolan, 'Prenatal Genetic Testing', *Pediatric Annals* 38/8 (2009), 426–30; E. Fragouli, 'Preimplantation Genetic Diagnosis: Present and Future', *Journal of Assisted Reproduction and Genetics* 24/6 (2007), 201–7.

53   F. Fukuyama, *Our Posthuman Future: Consequences of the Biotechnology Revolution* (New York, NY: Picador, 2002); G. Stock, *Redesigning Humans: Our Inevitable Genetic Future* (New York, NY: Houghton Mifflin Company, 2002). NY: Picador, 2002.

While genetic determinism is usually encountered in this hubristic context, surprisingly it is also encountered in quite a different context. This is in the argument that personal human life commences at fertilization, where personhood or 'ensoulment' is determined by genetic makeup.[54] In commenting on this position, Deane-Drummond writes, 'This downplays the importance of the environment, and of implantation in particular, as essential for human formation, equating the start of development with conception misses the point here.'[55] While this position may not set out to advance genetic determinism as usually encountered, it has clear overtones of this way of thinking. It is as deterministic in its rationale as are any of the hubristic forms. In other words, genetic determinism has crept into the thought forms of theologically based writers.

Up to this juncture I have confined my attention to genetic determinism alone, as though only genes matter. This is what may be referred to as one-factor determinism. However, this is not the end of deterministic-based approaches. In his very helpful analysis of genetic determinism, Ted Peters takes as his starting point two-factor determinism, genes plus the environment.[56] This is to make the point that while this differs markedly from one-factor determinism, genes alone, it may still be deterministic. Taking the example above of the MAO-A gene, and the presence or absence of abuse in childhood, one can accept both factors as precipitating causes within a deterministic paradigm. When both are present, aggression is inevitable or at least highly likely. This approach leaves no room for human freedom. But what is human freedom and where does it fit in? Is it a vitalistic or emergent force that is somehow introduced into a mechanistic system? Alternatively, is human freedom entirely explained by appeal to genes and the environment?

Of the many issues raised by these questions an important distinction is that between the whole and the parts, the individual as a functioning whole

---

54 D. A. Jones, *The Soul of the Embryo: An Inquiry into the Status of the Human Embryo in Christian Tradition* (London: Continuum, 2004).

55 Celia Deane-Drummond, *Genetics and Christian Ethics* (Cambridge: Cambridge University Press, 2006), 145.

56 Peters, *Playing God?*

and the cellular components making up that whole. The tension within scientific descriptions arises from the drive to concentrate on the cellular or subcellular (the genetic level in this case) to the exclusion of how these fit together as a system. It is in dissecting and reducing to manageable isolates that science achieves the understanding it does; without this process there would be no genetics. However, an accumulation of this knowledge by itself will not ensure an end-result that explains how the system operates, let alone the nature of human freedom. Peters asks, 'If the genes have been our puppeteer, did they themselves give us what we need now to cut the strings and take control? Is human freedom itself a gift of the genes? If not, then what is the condition that makes possible transcendence of our genetic inheritance?'.[57]

He argues that genetic determinism as human determinism is unsupportable scientifically. For him it is a category mistake in that freedom is found at the level of the whole person and not at the level of its component parts. Secondly, as already discussed above, genes at the molecular level are not determinative in any absolute sense, while mutations are the result of chance. Thirdly, human freedom could itself be the product of genetic determinism.

This raises the issue of whether there is a third factor that makes us free, something other than either genes or the environment. The problem with this is to determine the nature of this third factor, and this is where most discussions falter or at least stop. Human agency has to come in at some juncture, a necessity clearly recognized. Reichenbach and Anderson comment, 'But what room then remains for human agency? [...] Without choice there cannot be moral responsibility. But choice means that persons contribute something to the action over and above what derives from their genetic heritage and environmental input'.[58]

But is human agency independent of the two or does it emerge from our genetic and possibly environmental backgrounds? Whatever answer one gives to this question, it will be far removed from the 'gene for X'

57   Peters, *Playing God?*, 30–1.
58   Reichenbach and Anderson, *On Behalf of God*, 269.

explanation. This is because it refuses to go down the road of genetic essentialism, by refusing to equate the human self with genes, and to reduce human beings to molecular entities. It eschews the gene myth even while being prepared to take seriously the contribution of genes to what we are as whole persons. Genes are not sacred and should never be bestowed with 'soul-like' qualities.[59] There is a clear message to emerge from this: do not place undue reliance on genetic inheritance in isolation of other contributory factors to what makes us moral beings; such reliance reflects a cultural myth rather than scientific data. This message applies equally to those who would view all behavioural traits (whether exemplary or evil) as genetically fashioned, and also to those who would support the inviolability of the embryo and fetus on the grounds of genetic predetermination. Human beings, whether prenatal or postnatal, are more than a conglomeration of genes. Environmental factors are integral to any view of human agency, but it is probably not prudent to be too deterministic in any of our thinking even when genes are viewed within an environmental context. To quote Peters again, 'There is definite continuity between our DNA and destiny. The future will not nullify our present or past. Our spirit will not escape our biology [...] theologically speaking [...] genetic determinism and environmental determinism combined have not yet brought us to our full humanity. We are still on the way so to speak'.[60]

Genetic lottery

It is evident from the above that much of the debate on genetics revolves around an apparent lack of choice: I will only be able to be and act in ways predetermined by my genetic make-up. That is the emphasis that has been

59   Dorothy Nelkin and M. Susan Lindee, *The DNA Mystique: The Gene as a Cultural Icon*, 2nd edn (Ann Arbor, MI: University of Michigan Press, 2004).
60   Peters, *Playing God?*, 63–4.

the central theme to date of this chapter. However, this is not the whole of the debate. It has already become apparent that the hard determinism of genetics is blunted when epigenetics and the environment are taken into account. But, as touched on previously in this chapter, the genetic clinic raises issues of choice, rather than lack of choice. Here choices are being made between one embryo and another, and hence one future individual and another, on the basis of their genetic status. This may appear to have overtones of the 'gene for X' mentality, but the clinical situation is associated with genetic diseases caused by a single dominant gene, such as cystic fibrosis (see chapter 1), Huntington's disease, Tay-Sachs disease and sickle cell anemia, plus over 300 other conditions.[61] In these instances, the gene in question will manifest its presence in the particular disease; there is a direct correlation between the two.

There is real choice in these cases, even if the way of avoiding the condition is by refraining from implanting the embryo with the gene in question. This raises the issue of deciding whether choice itself is appropriate. Should the prospects opened up by genetic advances be welcomed or should we accept the genes with which we are endowed? This is where the notion of the genetic lottery enters the equation, according to which the genetic constitution with which individuals are endowed is a random mix of genes coming from their two parents. This lends itself to the view that people have to live with whatever they are given by nature.[62] According to this, choices should not be made between genes, accepting some but not others; in practice this rules out any choice between one embryo and another following genetic testing.

The intriguing aspect of this stance is that those who wish to adhere to the 'natural' have to advocate a form of genetic determinism. Whatever

61   In the UK before PGD clinics are permitted to test for a condition or combination of conditions, the HFEA must first agree that the condition they want to test for is sufficiently serious. There is a list of conditions that the HFEA has so far agreed that it is acceptable for clinics to use PGD to test for. See Human Fertilisation and Embryology Authority, 'PGD Conditions Licensed by the HFEA', <http://www.hfea.gov.uk/cps/hfea/gen/pgd-screening.htm> accessed 18 June 2013.

62   M. J. Sandel, 'The Case against Perfection', *The Atlantic Monthly* 293 (2004), 51–62.

genetic mix is inherited is the mix that should be adhered to, thereby reject-ing the choice option. Christians may express the genetic lottery in terms of God's will, so that choosing one embryo over another challenges this. Frequently, however, this is not the way in which the choice is phrased; rather, it is expressed in terms of the value placed on embryos. If embryos are regarded as inviolable they cannot be rejected. The result is that genetic issues are rendered irrelevant. This is problematic, since the question of genetic choice is a profoundly practical one, presenting patients and their families with real choices that have to be made when confronted by children or future children with certain genetic conditions. But should patients be faced with these choices? Should scientists have undertaken studies that have ended up with making such choices available, and should clinicians provide them in clinics? Or would it have been preferable if none of this research had occurred and everyone remained in ignorance?

It is strange that questions like this can be asked, when the knowledge available will contribute directly or indirectly to the betterment of the health of future generations. Admittedly, all is not straightforward either ethically or theologically, but there is no hint here of producing dramati-cally different beings, or of modifying them according to outrageously new designs. This is therapy in a fairly mundane sense, even if technologically sophisticated by most current standards. It is true that some commentators consider that the battle for the human 'soul' has already been lost, and they envisage the human race hurtling towards a posthuman, technologically driven future, with the propensity to subvert human values by creating separate classes of enhanced and unenhanced human beings.[63] Dramatic as the language is, such prospects are light years removed from the reality of the counselling room.

Such a stance, if interpreted simplistically, leads to the complemen-tary stance that is prepared to accept whatever genetic conditions emerge. However, the fabric of the human body incorporates genetic variables leading to disastrous disease states that have traditionally been coped with according to the capabilities of the medical knowledge of the day. Those

---

63   F. Fukuyama, *Our Posthuman Future.*

who argue against genetic choice are dispensing with this medical tradition by ignoring the prospects opened up by genetic-based approaches. However, as Peterson has argued, 'Genes create terrain and not destiny. A good genetic start does not guarantee a good outcome, it just makes such more likely. As in most of life, one does not have guarantees, but one can improve the likelihood of a more capable life'.[64] Elsewhere Peterson, in assessing how Christians are to approach medical technology more generally, concludes that, since as God's people we are being created, redeemed and transformed by God, we are to participate in processes aimed at 'sustaining, restoring, and improving what has been temporarily entrusted to us'.[65]

The heart of the debate is to determine what constitutes sustenance, restoration and improvement, especially when decision-making may involve the destruction of embryos and even of fetuses. The latter takes us out of the genetic realm, and so is not strictly relevant in the context of the genetic lottery. However, the thrust of the debate has not changed, since the only difference is when during gestation congenital anomalies can be detected in the clinic. It is true that aspects of the ethical debate will have changed, non-implantation of embryos versus the abortion of fetuses, and therefore the growing relationship with the mother and the network of other relationships are quite different.

One of the most contentious and high profile conditions is that of Down Syndrome, detection of which prenatally leads to very high rates of abortion.[66] However, the attributes of those with Down Syndrome are fiercely defended by individuals and organizations that see in these individuals special and precious characteristics.[67] While this is not the place

---

64   Peterson, *Changing Human Nature*, 183.

65   James C. Peterson, *Genetic Turning Points: The Ethics of Human Genetic Intervention* (Grand Rapids, MI: Wm. B. Eerdmans, 2001), 89.

66   Robert Cole and D. Gareth Jones, 'Testing Times: Do New Prenatal Tests Signal the End of Down Syndrome?', *New Zealand Medical Journal* 126/1370 (2013), 96–102.

67   J. Swinton and Brock B., eds, *Theology, Disability and the New Genetics: Why Science Needs the Church* (London: T&T Clark, 2007); A. Yong, *Theology and Down Syndrome: Reimagining Disability in Late Modernity* (Waco, TX: Baylor University Press, 2007).

to trace the dissension over Down Syndrome, some of the defenders of those with this condition contend that they are differently able for a substantial reason, and that this condition is part of God's plan for a diverse community.[68] May it be then that disability is a blessing, in response to which Peterson questions why attempts are regularly made to treat other comparable conditions.[69]

Down Syndrome, in particular, brings to light contrasting perspectives: those committed to decreasing the burden of ill health and disability in society, and those committed to looking after and appreciating the disabled and all they bring to the human condition. Both perspectives have to be listened to, and care has to be exercised that we do not rush to applaud either the technological imperative or a stance that equates disability with God's plan for human kind. The genetic lottery will not disappear and we have to learn how best to tackle it and live with its consequences. I see no place for passive acceptance of all it brings.

Genes and the person

The world of behavioural genetics points clearly to the conclusion that aspects of our character and personal identity have a genetic basis. This is not surprising, since our bodies are integral to who we are as people. Genetic factors are inevitably involved, even at the deepest levels of what makes us the people we are. But this in no way threatens the conception of a person as a rational being, capable of taking responsibility for oneself as a free agent. Neither does it detract from our ability to act as God's agents and stewards in his created order.

Human beings have a limited freedom, one constrained by their biological and environmental circumstances and also by their genetic make-up.

68 Yong, *Theology and Down Syndrome*.
69 Peterson, *Genetic Turning Points*.

They are not perfectly free. Through this self-understanding we can begin to appreciate our moral and spiritual limits, as well as our addictions and predispositions. We may also begin to see how God's grace can renew what we are as people, including possibly the ways in which genes are expressed in our body systems.[70]

What is of crucial significance is the ability to be oneself and to relate productively to others within the human community. Relationships such as these emanate from our personhood, and Christians would say as those made in the image of a triune God. The manner in which humans are treated should always be viewed within the broader context provided by human relationships, and never simply within the much narrower framework of biological parameters. Any choices we make should be choices to benefit people, and not simply to enhance disconnected building blocks, whether genes, brains or livers.

To argue like this is to argue for a person-centered model, and this is the model that governs every facet of my approach to genetics. I regard it as a model that is consonant with biblical and theological imperatives. We are people made in the image of a personal triune God, and created to relate to each other within community as well as to God as creator and redeemer.

In terms of this person-centered model, it is acknowledged that we make choices for ourselves and on behalf of others, because people have to make choices. Some of these choices will not raise any genetic or techno-logical issues and do not generally elicit vigorous ethical debate. However, others will, such as when genetic choices are made at the earliest stages of an individual's existence – probably when they were embryos. Once the notion of choice is raised in a genetic context, it introduces the possibil-ity of design – choosing and, to a very limited extent, designing people.[71] However, any design within genetics is of a far more limited and humble variety than so often encountered in these debates. It is far removed from

---

70 R. Cole-Turner, 'Soma, Psyche, Sin, and Salvation: Exploring the Relationship between Genetics and Theology', in M. L. Y. Chan and R. Chia, eds, *Beyond Determinism and Reductionism* (Adelaide: ATF Press, 2003), 16–35.

71 Jones, *Designers of the Future.*

the bravado and hubris associated with the picture of a factory production line of identical and preordained babies. The challenge is to determine how we do these things, and under what circumstances we do them, because this is where responsibility, judgement, and discernment come into play. We cannot do anything we like; we should not wish to be able to do anything we like. But we should do all we can to improve the quality of the lives of those around us, whether by using biological means or simply by treating them as beings of importance and as people who matter.

Humility is essential for rigorously assessing the merits of what can and cannot be accomplished by genetic science. Using a therapeutic and person-centered framework, our eyes can be directed towards what can realistically be accomplished to the benefit of patients. This is a far cry from the hubris sometimes encountered, but also from the anti-hubris that has become so caught up in the fear of extravagant claims that it has lost sight of the good that can be accomplished by utilizing some of these technologies.

Nevertheless, caution is to be exercised and this is to beware of obsession with the normal, something that could be accentuated by any of the current biomedical technologies. The genetic realm is as limited as any other, and programs to design wonderful new human beings are futile. On the other hand, the rejection of a modicum of limited and very cautious design is the outcome of a spirit of fear rather than a spirit of faithfulness. We are to do what is consistent with the nature and purposes of God, and are to assess all scientific developments by the benchmark of whether they appear to promote God's work in creation. Daunting as these tasks are, and inadequate as we may be to tackle them, they are enriched when theological, scientific and ethical insights are brought to bear on them in an integrated fashion.

Discussion of topics like choosing children's genes tends to revolve around choosing genes for fair hair, blue eyes, intelligence, physique, good looks, avoiding baldness or whatever. The ephemeral nature of these longings only serves to demonstrate their superficiality, let alone the scientific precision, clinical complexities and expensive resources that would be required to achieve them. Unfortunately, instead of demythologizing such fantasies as empty claims (the gene myths that they are in reality), they are taken seriously and are used to construct tirades against realistic and therapeutically based genetic choice. The latter can then be dismissed on the ground that its goal is that of producing perfect babies, designed to order. These twin themes of perfectibility and designer babies carry powerful negative theological overtones, with their message that science is assuming redemptive powers; salvation can be found in biological manipulation, and the hope of a better life emanates from genetic intervention.[72]

The trouble is that these paradigms are based on little more than irresponsible hype – scientific as well as journalistic. Christians rightly reject any such paradigms grounded in such quasi-scientific aspirations.[73] However, the Christian task should be that of debunking this fatuous mythology, and not using it to frighten and mislead. To use it as the foundation on which to construct a case against genetic intervention in the name of Christian imperatives is to fall into the same trap as those who look for a biological version of the new heavens and new earth. While the intentions of these two groups are radically different, they both accept the hubris implicit within a scientific vision that assumes that nothing lies outside its manipulatory abilities.

Starting from a baseline like this, any assessment (Christian or otherwise) of the prospects opened up by genetic intervention, will be mired in opposition to them. The rationale of this opposition is rejection of extravagant claims rather than an analysis of the prospects opened up by serious genetic science. Rejection of hubris (valid as it may be as a general principle) should not be a theological starting point. Far more relevant in this context is the embrace of humility – to enable a rigorous assessment of the merits of what can and cannot be accomplished by genetic science. Using a therapeutic framework, attention can be directed towards what can be realistically accomplished to benefit the patient. The good of the patient becomes the guiding principle; embedded within this is a commitment to improve the quality of the patient's life or to replace illness by health. This is both a positive hope and a realistic one. The genetic intervention

72   G. Stock, *Choosing Our Children's Genes: Redesigning Humans* (London: Profile Books, 2002).
73   Peterson, *Genetic Turning Points*.

may not work; our hopes may be dashed. But the attempt is to be encouraged as long as our expectations are guided by realistic clinical and scientific goals. There is no hint here of perfection or of ageless existence in a disease-free body. The dominant value is that of caring for those in need, and of utilizing powerful technologies in the service of those potentially capable of benefitting from them.[74] While the dividing line between therapy and enhancement is both unclear and shifting (see chapters 2 and 6), an emphasis upon the good of the person helps to keep the focus on what is largely a therapeutic agenda.

74   R. Cole-Turner, *The New Genesis: Theology and the Genetic Revolution* (Louisville, KY: Westminster/John Knox Press, 1993).

CHAPTER 5

Modifying brains and changing who we are

Brains and persons: from dualism to physicalism

My training is in neuroscience and hence it is not surprising that I am wedded to the centrality of the brain for so much of that which makes us human. This immediately sets me on a collision course with some Christian commentators whose starting point is provided by supposedly traditional Christian anthropology with its duality of body and soul, or even the three-part division of body, soul and spirit.[1] I do not see it as my task to argue against the validity or otherwise of such positions on the basis of the theological literature. Rather I wish to explore where a serious engagement with neuroscientific investigations will lead within a Christian framework. This is far from a new endeavour although its character has changed with the changing face of neuroscience.

Underlying this endeavour is a presumption that science is to be taken seriously, and that an understanding of brain function should inform any approach to the human person. This includes the approach of Christians. Once again, there should be no great surprise here, since people in general work on the premise that an understanding of the gastrointestinal tract, the liver or the kidneys is central to an appreciation of numerous aspects of health and disease. Mental health and incapacity is not any different, as was amply demonstrated in chapter 2.

---

1   J. W. Cooper, *Body, Soul and Life Everlasting: Biblical Anthropology and the Monism-Dualism Debate* (Grand Rapids, MI: Eerdmans, 1989); J. P. Moreland and S. B. Rae, *Body and Soul: Human Nature and the Crisis in Ethics* (Downers Grove, IL: Inter-Varsity Press, 2000).

This is readily accepted at the applied level, but what about the theological level when the status and nature of the human person is at stake?
To what extent, if any, can the person be reduced to a material organ, the
brain, with its constituent neurons, synapses, networks of neuronal processes, fibre tracts and brain regions specialized for performing different
functions? The impression is readily given that if we move too far in this
direction, we end up reducing human beings to nothing more than these
component parts. Perhaps the extreme reductionist position espoused by
Francis Crick's famous 'astonishing hypothesis' becomes the inevitable end
point. According to this, all human joys, sorrows, memories, ambitions,
sense of personal identity and free will 'are [...] no more than the behavior
of a vast assembly of nerve cells and their associated molecules.'[2] Bleak as
such a picture may appear it is an outworking of Crick's belief that 'The aim
of science is to explain all aspects of the behavior of our brains, including
that of musicians, mystics, and mathematicians.'[3] I would also contend
that it stands as an epitaph to neural determinism in its starkest form. But
is there any way of avoiding this harsh end point once the centrality of
brain processes is allowed sway?

It is in attempting to answer this question that the attractions of dualism become apparent. Why not allow the material brain its role, but ensure
that it is balanced by a non-material soul or mind? The mind-body problem is an issue within philosophy as old as philosophy itself, and it is only
in recent years that neuroscience has intruded into the debate, and in so
doing has altered its character. This in turn has opened debate between
philosophers and neuroscientists concerning the role, if any, that the latter
has in it.[4] Inevitably, theologians also recognize the stake they have in the
debate, although their emphasis is on the soul rather than the mind,[5] and

2   Francis Crick, *The Astonishing Hypothesis* (London: Touchstone Books, 1994), 3.
3   Crick, *The Astonishing Hypothesis*, 259.
4   Patricia S. Churchland, *Neurophilosophy: Toward a Unified Science of the Mind-Brain*
     (Cambridge, MA: MIT Press, 1986); Paul M. Churchland, *The Engine of Reason,
     the Seat of the Soul* (Cambridge, MA: MIT Press, 1995).
5   See Cooper, *Body, Soul and Life Everlasting*; Joel B. Green and Stuart L. Palmer, eds,
     In Search of the Soul (Downers Grove, IL: Inter-Varsity Press, 2005); William Hasker,

this has led to a bewildering array of perspectives. These include substance dualism, holistic dualism, emergent dualism, and nonreductive dualism (physicalism). A major figure in neuroscience who espoused a form of radical dualism or interactionism was Nobel laureate Sir John Eccles.[6] In the preface to *The Human Psyche*, based on his Gifford Lectures, Eccles wrote: 'The aim has been to demonstrate the great explanatory power of dualism-interactionism in contrast to the poverty and inadequacy of all varieties of the materialist theories of the mind.'[7]

The array of theological approaches to dualism show that it is unlikely that this approach is the sole bastion against materialism, although they post-date Eccles's contribution and also do not attempt to go into the neuroscientific minutiae that was his forte. While Eccles's approach has not garnered substantial support over subsequent years, it stands as testimony to an approach built firmly upon a neuroscientific base, but owing a great deal to philosophical assumptions, notably, the primacy of consciousness. For Eccles and Popper, their defence of human dignity and meaning rests on an explicit dualism between the self and the brain with the self-conscious mind acting on neural centres in the brain and so modifying the dynamic spatio-temporal patterns of neural events. This is worked out in considerable detail using Popper's three-world-view (matter and energy, self-awareness, culture). Interaction can take place between these 'worlds', and for Eccles this enables the self-conscious mind to influence neural events in special areas of the neocortex.[8] The precise region/area of the brain postulated to be involved was always highly speculative in Eccles's thinking and hence the details are not of concern. What does emerge is a

---

The Emergent Self (Ithaca, NY: Cornell University Press, 1999); Moreland and Rae, *Body and Soul*; Nancey Murphy, *Bodies and Souls, or Spirited Bodies?* (Cambridge, MA: Cambridge University Press, 2006).

6   See J. C. Eccles, *The Human Mystery* (Berlin: Springer, 1979); J. C. Eccles, *The Human Psyche* (Berlin: Springer, 1980); K. R. Popper and J. C. Eccles, *The Self and Its Brain: An Argument for Interactionism* (Berlin: Springer, 1977).

7   Eccles, *The Human Psyche*, vii.

8   D. Gareth Jones, *Our Fragile Brains: A Christian Perspective on Brain Research* (Downers Grove, IL: Inter-Varsity Press, 1980).

fundamental problem of dualism and this is the feasibility of one sort of substance acting on another sort of substance. Is this feasible, or does it imply that if the mind/self/soul acts on modules of neurons, the mind/self/soul is itself a module? Alternatively if the mind/self/soul acts in some inexplicable way, does this make it an inexplicable entity?[9]

Even more important is the observation that, while Eccles emphasized our awareness of our conscious selves, this tended to be lost in the detailed outworking of the neuroscience. While this may not have been intentional, the desire to distinguish the material brain from an immaterial mind/self/soul misses the intimate relationship we as people have to our bodies.[10]

Utilizing quite a different vantage point, Donald MacKay provided an alternative perspective of a Christian working in the neurosciences.[11] He described his position as comprehensive realism.[12] For him we have to think of human beings as a unity, an indivisible whole, but capable of being described in two different ways: the I-story (Inside story) and the O-story (Outside story; also referred to as the brain-story), corresponding respectively to what we know as conscious agents and how we can be described in terms of our brains. Both are equally real aspects of what makes us what we are, and are complementary. MacKay writes: 'They are not exactly symmetrical, since there can be some changes of brain-state without any change of conscious experience, whereas the converse [...] is not the case [...] each of them stands to be reckoned with at its own level and must be reckoned with if we are not to miss some of the reality that is man'.[13]

Rather than move in the direction of epiphenomenalism, MacKay (1988) argued for the logical complementarity of the I-story and the O-story (brain-story), pointing to a duality of aspects rather than of substances. On

---

9   Jones, *Our Fragile Brains*, 260–1.
10   Jones, *Our Fragile Brains*.
11   Donald M. MacKay, *Brains, Machines and Persons* (London: Collins, 1980); Donald M. MacKay, *Human Science and Human Dignity* (London: Hodder and Stoughton, 1979); Donald M. MacKay, *The Open Mind and Other Essays*. Edited by M. Tinker (Leicester: Inter-Varsity Press, 1988).
12   MacKay, *Brains, Machines and Persons*.
13   MacKay, *Brains, Machines and Persons*, 83.

this basis he contended that 'there are no rational grounds for taking the material aspect as more "real" than the mental, and still less for identifying the two and speaking of thinking and deciding as attributes of matter'.[14] Further, he was adamant that Christian perspectives start from the dependence of physical events on God and that this offers a positive incentive for Christians to take seriously the physical aspect of human nature and therefore neural events.[15] This equal status accorded to the primacy of consciousness (I-story) and brain activity (O-story) veers markedly from that of Crick for whom 'I' can only be understood in terms of neural activity.[16]

After surveying the neuroscientific and theological literature, psychologist Malcolm Jeeves concluded that:

> [S]elf-conscious human agency [...] is the primary ground with which we have to do, rather than mind or matter. We may project this concept of human agency onto the outside world in terms of an image of brain events or may take the standpoint of the agents themselves experiencing mental events [...] There is here then an intrinsic duality about the reality with which we have to deal, but this does not need to be taken to the extreme of dualism.[17]

It appears, therefore, that when serious attention is paid to neuroscience, there is no escape from incorporating it into a view of personhood. However, neuroscience by itself is not the only or even the determinative factor in moving in this direction. In his assessment of what it means to be human, theologian Joel Green argues that the Genesis narrative does not provide a 'basis for the existence of an ontologically distinctive entity known as the "soul" [...] nor with the identification of a person's true "self"'

---

14   Donald M. MacKay, 'Brain Science and the Soul', in R. L. Gregory, ed. *The Oxford Companion to the Mind* (Oxford: Oxford University Press, 1987), 723–5, 724.

15   MacKay, *Brains, Machines and Persons.*

16   Crick, *The Astonishing Hypothesis.*

17   Malcolm Jeeves, 'Human Nature: An Integrated Picture', in Joel B. Green, ed. *What About the Soul?: Neuroscience and Christian Anthropology* (Nashville, TN: Abingdon, 2004), 171–89, 177.

with such an entity'.[18] For him both theology and neuroscience highlight the embodiedness and relationality of human beings. This, in turn, leads to awareness that there are neurobiological correlates of our actions, attitudes, patterns of behaviour, beliefs, moral formation and reformation. Even talk of being a new creation in Christ (2 Cor. 5:16–17) will, one imagines, be reflected in the brain. This is what Green refers to as 'salvation-as-embodied transformation'.[19] Hence it can be argued that 'when we seek to construct a neurobiological portrait of the human person, we are provided with a framework that is entirely consistent with Christian emphases on the wholeness and coherence of the person'.[20]

Against this background it appears that, while a scientific study of the nervous system will not by itself tell us all we want to know about people and their functioning, a barely functioning brain or a severely damaged one will have major consequences for the individual and their relationships.[21] As outlined in the following section, extreme damage to the brain demonstrates that there is an intimate relationship between the brain and the person, even if this is not the only factor to be taken into account. Care has to be taken to avoid the pitfall of thinking that we are no more than brains ensheathed in bodies; the notion that we are embodied persons sits far more comfortably with the Christian claim that we are created in the image of God.

Implicit within the positions espoused by Mackay, Green and Jeeves is rejection of the notion of a 'soul' as conventionally conceived, namely, that of an immaterial entity that in some way characterizes human beings. This is still hotly debated within theological circles, as evidenced by writers such as Cooper, and also Moreland and Rae who consider that the person is more than the material components or genetic make-up of the body, and

---

18   Joel B. Green, *Body, Soul, and Human Life: The Nature of Humanity in the Bible* (Grand Rapids MI: Baker Academic, 2008), 70.

19   Green, *Body, Soul, and Human Life*, 139.

20   D. Gareth Jones, 'A Neurobiological Portrait of the Human Person: Finding a Context for Approaching the Brain', in Joel B. Green, ed. *What About the Soul? Neuroscience and Christian Anthropology* (Nashville, TN: Abingdon Press, 2004), 31–46, 46.

21   Jones, 'A Neurobiological Portrait of the Human Person'.

is to be identified with an immaterial soul.[22] For them the soul governs the law-like development of the body and enables the person to maintain its identity from conception onwards.[23] Very occasionally, as we saw above, neuroscientists (such as Eccles), also adhere to this view of a soul.

For Christian philosopher Nancey Murphy, the New Testament writers attest that humans are psychophysical unities and that the Christian hope for eternal life depends on the reality of bodily resurrection and not on the ongoing existence of an immortal soul.[24] Additionally, what is central to human existence is the network of relationships with others and also with God. However, she does not consider that study of the biblical writers leads definitively to either physicalism or dualism.[25] Murphy herself argues for a physicalist account of the person, since humans are purely physical organisms, although for her a satisfactory resolution lies in what she calls 'nonreductive physicalism'. By this she means that, if there is no soul, one has to look to brain function to explain higher brain capacities, although a full explanation will also take account of human social relations, cultural factors and most importantly God's action in the lives of human beings.[26]

In light of these considerations, it is imperative that serious account is taken of the functioning of the brain, both in health and disease. Working from a physicalist position, I accept that brain damage will on occasion have profound effects on an individual's personality and value systems, and that these changes have to be incorporated into any Christian framework. Interestingly, similar challenges have to be faced by dualists of all persuasions.

---

22   Cooper, *Body, Soul and Life Everlasting*; Moreland and Rae, *Body and Soul*.
23   Moreland and Rae, *Body and Soul*, 343.
24   Murphy, *Bodies and Souls, or Spirited Bodies?*, 22. See also Oscar Cullmann, 'Immortality of the Soul or Resurrection of the Dead? The Witness of the New Testament', in Terence Penelhum, ed. *Immortality* (Belmont, CA: Wadsworth, 1973), 53–85; Wolfhart Pannenberg, *Jesus – God and Man*, Translated by Lewis L. Wilkins and Duane A. Priebe (London: SCM Press, 1968), 87; N. T. Wright, *The Resurrection of the Son of God: Christian Origins and the Question of God* (London: SPCK, 2003); Wright, *Surprised by Hope*.
25   Murphy, *Bodies and Souls, or Spirited Bodies?*, 37.
26   Murphy, *Bodies and Souls, or Spirited Bodies?*, 116.

Plasticity of the brain

The previous discussion suggests that a neuroscientific perspective does not lead inevitably to a machine-type model. This is because the brain is the antithesis of a machine, since it is highly responsive to numerous environmental stimuli at all stages of life. It is nothing like a machine with its fixity and rigidity. Such a model is misleading not because it fails to do justice to Christian aspirations, but to scientific ones. The two-way interactions between an individual's brain, and the worlds internal and external to that individual, point to the richness of its multi-dimensional context. In no sense can the brain be isolated from an individual, in the way in which the heart can be; neither could it be replaced by another brain without destroying the integrity of the individual as the person he or she is known to be.

These possibilities stem from a basic neural phenomenon, namely, the plasticity of the brain,[27] and even the production of new neurons in the adult brain.[28] What we are as persons is not laid down once and for all in the genome. While it is true that the basic ground plan for an individual's brain is specified in the genome, the detailed patterns of synaptic connections that link its innumerable neurons are fashioned by a host of influences throughout life.

It is because of this close interrelationship between the developing brain and extrinsic factors, that normal development can readily be disrupted. Changes to the environment, such as delays to the arrival of sensory impulses during development, result in a different brain organization from what would have been the case under other conditions.[29] Consequently,

27   A. Pascual-Leone et al., 'The Plastic Human Brain Cortex', *Annual Review of Neuroscience* 28 (2005), 377–401.
28   P. Rakic, 'Neurogenesis in Adult Primate Neocortex: An Evaluation of the Evidence', *Nature Reviews: Neuroscience* 3/1 (2002), 65–71.
29   J. A. Markham and W. T. Greenough, 'Experience-Driven Brain Plasticity: Beyond the Synapse', *Neuron Glia Biology* 1/4 (2004), 351–63; H. Neville and D. Bavelier, 'Human Brain Plasticity: Evidence from Sensory Deprivation and Altered Language Experience', *Progress in Brain Research* 138 (2002), 177–88.

numerous influences during development can have devastating consequences for a child's subsequent intelligence and behaviour. The converse also holds: a stimulating environment can increase the complexity of neural organization, in turn altering many facets of the behavioural repertoire of the growing individual. While the same phenomenon does not continue unabated into adult life, elements of plasticity are still found.

It is hardly surprising, then, that it is the specific fine-tuning of the synaptic connections between neurons in any given brain that contributes substantially to that individual's uniqueness and personhood. What we are as people emerges from an ongoing dialogue between the neural and genetic material we inherit, the worlds we occupy and the worlds we ourselves construct.[30]

Extensive plasticity like this makes possible the enormous range of human beings' intellectual abilities and spiritual gifts, their personalities and their creative abilities. It is on account of these neural capabilities that we can be human persons epitomized by responsibility, creativity and responsiveness to each other and to God. Remove neural plasticity and many of the marks of true human personhood disappear. Our biological uniqueness as individuals mirrors our theological uniqueness as persons created by God. The relevance of this for our present discussion is that so much of our biological uniqueness stems from the characteristics of our brains, characteristics that in turn emanate from our genetic inheritance, numerous facets of our pre- and postnatal environment, our responses to the world we encounter, the attitudes we display towards our fellow human beings, and our response to God.

---

30   J. P. Curley et al., 'Social Influences on Neurobiology and Behavior: Epigenetic Effects During Development', *Psychoneuroendocrinology* 36/3 (2011), 352–71; J. Stiles, 'Brain Development and the Nature Versus Nurture Debate', *Progress in Brain Research* 189 (2011), 3–22.

Fragile brains and distorted persons

The extent of the interdependence between the brain and person is demonstrated by the way in which pathologies of the brain can have devastating consequences for the integrity and wholeness of a person. Consider the following extreme examples.

> Chris Birch worked in a bank, was planning a wedding with his girlfriend, and spent his average Saturday watching football with his mates. But one day he had an accident and suffered a stroke that completely changed him. He left his job, grew to hate sport and realized he was gay. His mother barely recognizes him and he is now a hairdresser. He himself says that he looks like a different person. 'I'm nothing like the old Chris now'.[31]

> Tommy McHugh nearly died after two blood vessels burst in the back of his head. After a week in a coma he woke up with an uncontrollable urge to create and began writing poetry, painting the interior of his home, sculpting and carving. Prior to this he had never painted or written poetry. In fact he had been a builder and handyman, and had also been in and out of young offenders institutions. In adult life he had had a dubious reputation as a street fighter. Now he is a compulsive artist. He says, 'I'm not an artist. I'm just letting the creativity flow out of me'. His friends don't know what to make of this new person, who often speaks in rhyme, loves kittens, and wants to know what life means.[32]

> One day Jon Sarkin a thirty-five-year-old chiropractor, was playing golf, when he suddenly started having symptoms due to a small blood vessel pressing on his auditory nerve. This was treated surgically and the result was a massive stroke. When he

---

31   L. Salkeld, 'Burly Rugby Player Has a Stroke after Freak Gym Accident ... Wakes up Gay and Becomes a Hairdresser', *Daily Mail* (9 November 2011) <http://www.dailymail.co.uk/health/article-2058921/Chris-Birch-stroke-Rugby-player-wakes-gay-freak-gym-accident.html> accessed 22 June 2013.

32   Daily Mail Reporter, 'Ex-Street Fighter, 60, Turned into a Fanatical Artist by a Brain Haemorrhage That Physically Altered His Mind', *Daily Mail* (16 March 2010) <http://www.dailymail.co.uk/news/article-1256967/Man-survived-brain-haemorrhage-transformed-fanatical-artist-paints-18-hours-day.html> accessed 22 June 2013.

eventually regained consciousness, he was a different person. Twenty-three years later he described it in these words: 'Beforehand, I knew who I was, more or less. But after this I didn't – and I still don't, not fully. Say you have a curve that gets closer and closer to another line without ever meeting. It's a logarithm. That's me. My sense of self is logarithmic'. Now at the age of sixty he is a well-established artist. He states quite categorically 'the old Jon is gone'.[33]

In all these cases the brain injuries may be disinhibiting the brain so that it makes connections that would not normally be made. The damage is generally to the frontal and temporal lobes that are associated with creativity. These are extreme examples of pathological states, and they are but the tip of the iceberg. Numerous examples of a host of diverse conditions could be brought forward to illustrate the same point: integrity of the brain and neural processes is essential for maintaining the features that characterize individuals, or as in these particular cases, the core features of what they are as persons.[34]

The problem in so many of these cases is that detailed accounts of the stages through which the individuals go when experiencing the brain damage and any subsequent recovery is lacking. Some of these gaps were filled in by a fascinating account of a stroke from the personal experience of a neuroscientist, Jill Bolte Taylor.[35] In 1996 when thirty-seven, this Harvard-trained neuroscientist suffered from a massive stroke in her left hemisphere. Within four hours she had lost most of the functions associated with normal existence – walking, talking, reading and writing. Over the eight years of her recovery she was able to reflect on the respective contributions of her left and right hemispheres. She came to appreciate the contribution of her right hemisphere with its intuitive and kinesthetic

---

33 R. Collin, 'The Curious Case of an Accidental Artist', *Telegraph* (1 August 2011) <http://www.telegraph.co.uk/health/healthnews/8670516/The-curious-case-of-an-accidental-artist.html> accessed 22 June 2013; A. E. Nutt, *Shadows Bright as Glass: The Extraordinary Transformation of One Man's Brain and the Neuroscience That Makes Us Who We Are* (Piatkus Books, 2011).

34 Norman Doidge, *The Brain That Changes Itself* (Melbourne: Scribe, 2010).

35 Jill Bolte Taylor, *My Stroke of Insight: A Brain Scientist's Personal Journey* (London: Hodder, 2006).

features as opposed to the far more logical and sequential features of the left hemisphere on which she had previously depended. Her reflections during these years are stunning as she grappled with the loss of left hemisphere functions.

In the first few days she recognized that she had 'died that morning and no longer existed [...] My entire self-concept shifted as I no longer perceived myself as a single, a solid, an entity with boundaries that separated me from the entities around me'.[36] Much later as she learned the basic skills, starting with reading and writing, she was aware of the conflict between the demands of her two hemispheres. As her left hemisphere skills increased she no longer found all her old personality traits acceptable since she had discovered a peace that she put down to the emergence of her right hemisphere skills. 'My stroke of insight is that at the core of my right hemisphere consciousness is a character that is directly connected to my feeling of deep inner peace. It is completely committed to the expression of peace, love, joy, and compassion in the world'.[37]

The trauma led to a transformation in Bolte Taylor's outlook on life, a change she interprets as a re-balancing in the differential influence of the two hemispheres. The crucial role played by the brain's plasticity also becomes crystal clear, as does the influence she was able to bring to bear on the manner of her recovery. In her picturesque language: 'Thanks to their neural plasticity, their ability to shift and change their connections with other cells, you and I walk the earth with the ability to be flexible in our thinking, adaptable to our environment, and capable of choosing who and how we want to be in the world'.[38] While there may be a hint of exaggeration in this statement, it makes the valuable point that, in exercising responsibility, our brains are changed.

All these are pathological instances, when something has gone drastically wrong. The individual is no longer what they were. The original has gone and has been replaced by someone with a different identity. Who is

36   Bolte Taylor, *My Stroke of Insight*, 68–9.
37   Bolte Taylor, *My Stroke of Insight*, 133.
38   Bolte Taylor, *My Stroke of Insight*, 176.

the real person, and with whom do we deal as the real person? What if the two have different values, let alone different interests?

Is it possible that the post-damage individuals are an improvement on the pre-damage ones, or is it the converse? Obsessive elements are pathological, even though creativity and artistic talent may be welcome. And what about the love of peace, calm and harmony as opposed to the drive to achieve and excel? There are no easy answers, except that care is required in drawing hasty conclusions without knowing how the post-damage individual relates to family, friends and colleagues and contributes to the lives and welfare of others. If the post-damage individuals are viewed as improved versions of the original they will be regarded as healthy; if viewed as damaged they will be viewed as in need of support and ongoing therapy. These considerations also apply to a Christian analysis, for which relationality is central.

Less extreme examples are also common. For instance, take patients with damage to their ventromedial prefrontal cortex and have impaired emotional responses and make aberrant, unusually utilitarian decisions when faced with a moral dilemma.[39] The significance of this is that it applies regardless of their moral or religious commitments prior to the injury.

Recent case studies on a unique individual with bilateral amygdala damage have revealed the role of the amygdala in mediating explicit responses to social and emotional events, in contrast to the prevailing conception of the amygdala as a primitive threat detector.[40] In particular, this patient fails to appreciate the appropriate interpersonal distance normally maintained by a sense of social comfort between individuals. While she can rationally comprehend others' sense of interpersonal space she simply does not feel the discomfort that too close proximity usually brings.[41]

---

39 M. Koenigs et al., 'Damage to the Prefrontal Cortex Increases Utilitarian Moral Judgements', *Nature* 446/7138 (2007), 908–11.

40 Rebecca M. Todd and Adam K. Anderson, 'Six Degrees of Separation: The Amygdala Regulates Social Behaviour and Perception', *Nature Neuroscience* 12/10 (2009), 1217–18.

41 D. P. Kennedy et al., 'Personal Space Regulation by the Human Amygdala', *Nature Neuroscience* 12/10 (2009), 1226–7.

While these are all examples of injuries it is possible to change values using drugs. For instance, some patients with Parkinson's disease have been transformed from law-abiding citizens into compulsive gamblers and obsessive pleasure-seekers as a result it appears of the dopamine enhancers they are receiving as treatment for the disease.[42] In this case, the patients have received too high levels of dopamine, a chemical that is seriously depleted in the disease. Lowering the dose results in a return to normal behaviour, so that in this situation there is a direct causal link between high levels of dopamine and compulsive behaviour.

There can be no doubt that a causal relationship exists between injury to certain brain regions and aberrant behaviour. Each of these (pathological) examples is a reminder of the intimate link between our physical brains and our standing as human persons. They are also a reminder of human vulnerability, in that any intrusion into the brain is an intrusion into the centre of what humans are as physical beings. A further example was encountered in chapter 4 where reference was made to the role of monoamine oxidase A (MAO-A), involved in regulating a neurotransmitter, serotonin, in the brain. An association between the MAO-A gene and aggressive behaviour has been found in one particular family with a high incidence of violence.[43] A subsequent study by other researchers also showed a link between the MAO-A gene and antisocial behaviour if the individuals concerned had also been mistreated and abused as children.[44]

A perspective amenable to Christian premises will assert that the ethical road is to ascertain the degree of moral responsibility within a framework of low MAO-A expression. The neural basis of thought and behaviour in no way threatens the conception of a person as a rational being, capable of taking responsibility for oneself as a free agent. Neither does it even hint that we cannot act as God's agents and stewards in his created order. It is up to us as persons to determine what we do with both our abilities and

42   M. L. Dodd et al., 'Pathological Gambling Caused by Drugs Used to Treat Parkinson Disease', *Archives of Neurology* 62/9 (2005), 1377–81.

43   Brunner et al., 'Abnormal Behavior Associated with a Point Mutation in the Structural Gene for Monoamine Oxidase A'.

44   Caspi et al., 'Role of Genotype in the Cycle of Violence in Maltreated Children'.

restrictions (no matter how neurally-based some of these may be). We are to use the resources at our disposal, rather than view ourselves as prisoners of our inheritance. The information provided by neural studies and behavioural genetics should be used to increase one's repertoire of understanding, so that one can come to terms with the behavioural conundrums with which we are all confronted. In the final analysis it is we who decide how we live and act, and what we believe. For some this freedom is severely restricted, due to developmental restrictions or later brain injury. But most are in a position to play a causal role in how to live and act.

There are no simple answers to those who give the impression of having become 'new persons' as a result of brain injury. In the extreme cases alluded to above there appear to be no ways of going back to the 'original person', even if this is desired. There are far too few instances recorded, let alone studied adequately, to make any significant generalizations, although similar predicaments arise with severe cases of Alzheimer's disease where the link between what was and what is now may be very different. However, even here there are connections and there can be little question that it is the pre-dementia person who provides the benchmark for what that individual represented and stood for. By contrast the stroke victims represent a more radical departure from a previous state to the present one. The obsessive nature of some of the reported behaviour points unequivocally to a pathological condition, so that the individuals should be approached within that context. Creative though some of these individuals undoubtedly are, they are also people to be viewed within a network of relationships that have ethical and theological significance.

This conclusion has been set in the context of brain damage, although it applies equally strongly where there is no pathology. There are strong connections between what individuals are as biological and physical beings, and how they live as people, since none of us can escape the repercussions of our biological make-up, but neither are the two identical. Most people are not prisoners of their biological make-up, but seek to live out what they are, with all their possibilities and limitations. And our lives are lived in terms of our worldviews and moral codes, whatever these may be. This perspective is useful when approaching issues as diverse as gender attraction and homosexuality, to alcoholism and eating disorders, where neural characteristics may play a role in making individuals what they are as physical entities. My argument is that this is an important part of the story but not the whole story.

Neural correlates and brain imaging

For many years it was possible to discuss brain issues in a highly theoretical manner, simply because little could be done to investigate what was actually happening in the brains of individuals, or little could be done to treat people with neurodegenerative conditions or following brain injuries. This is now changing on both counts, albeit gradually in the treatment area. However, there is a proviso.

The simple act of finding neural correlates for certain behaviours or attitudes provides few, if any insights into causative factors. Even if a certain brain structure is found to be strongly associated with religious experience, this says nothing about whether the structure generates that experience. Simply because brain region 'R' is active when behaviour 'B' is undertaken does not mean that changes in 'R' cause 'B' to take place. The opposite, in fact, could be the case, in that when an individual displays behaviour 'B' brain region 'R' is modified, and if this occurs sufficiently often, there are significant changes to 'R'. Yet again, the interplay between 'R' and 'B' may be so close that the only tenable conclusion is that there is no definitive causative factor. And yet the neural correlates detectable by brain imaging may question some aspects of our moral geography.

The existing literature points towards the ability to detect the neural correlates of an increasingly wide array of conditions and traits.[45] These include conscious and unconscious racial attitudes, conscious self-regulation of emotion, a range of personality traits, personality disorders and psychopathic conditions, serious criminal tendencies, drug abuse such as

45   Jones and Whitaker, *Speaking for the Dead.*

cocaine craving, preferences for products such as well-known drinks, and the decision-making process itself. All these in their different ways are illustrations of 'brain reading'.[46] These raise issues of considerable significance for society, while for Christian thinkers they serve as a stark reminder of the centrality of the brain for all models of the human person. In their own way they throw light onto the various models spanning dualism and physicalism alluded to earlier in this chapter.

Theoretically, it may prove possible to pinpoint the parts of someone's brain that are active whatever decisions are taken by that individual, including moral and spiritual choices. While a great deal of space in the neuroscientific literature is devoted to the relationship between neural functioning and aggressive behaviour, one can imagine that there is also a relationship between neural functioning and forgiving behaviour. The same comments would probably apply to the act of praying and it has been suggested that different types of prayer may be associated with different brain regions.[47] Changes in cerebral activity during glossolalia ('speaking in tongues') have been assessed using SPECT (single-photon emission computed tomography), an imaging technique less disruptive to the subject than fMRI (functional magnetic resonance imaging).[48] When compared to a religious state involving singing in English, subjects exhibited decreased activation in the prefrontal cortices, consistent with their description of glossolalia as non-voluntary. The scans also indicated decreased activation of the left caudate nucleus and a change in thalamic lateralization, which could be associated with the subject's altered emotional state.

There has been considerable interest in locating the brain regions involved in religious or spiritual experiences. Changes in cerebral blood

46   J. D. Haynes and G. Rees, 'Decoding Mental States from Brain Activity in Humans', *Nature Reviews Neuroscience* 7/7 (2006), 523–34.

47   U. Schjodt et al., 'Rewarding Prayers', *Neuroscience Letters* 443/3 (2008), 165–8.

48   J. J. McGraw, 'Tongues of Men and Angels: Assessing the Neural Correlates of Glossolalia', in D. Cave and R. S. Norris, eds, *The Body and Religion: Modern Science and the Construction of Religious Meaning* (Leiden: Brill, 2012), 57–79; A. B. Newberg et al., 'The Measurement of Regional Cerebral Blood Flow During Glossolalia: A Preliminary SPECT Study', *Psychiatry Research: Neuroimaging* 148/1 (2006), 67–71.

flow using SPECT have been studied during various types of meditation.[49] Both Franciscan nuns engaging in meditative prayer and Tibetan Buddhists performing visualisation meditation showed increased activity in the prefrontal cortex. In many regards, this is to be anticipated in terms of what is known about the functions of this part of the brain. One would like to know whether this in itself helps us understand more about the practice of meditation, and even whether it is to be encouraged as a religious ritual. Or is such neuroscientific knowledge irrelevant in religious terms? One answer put forward is that activation of the frontal lobes can help explain the intrinsically rewarding nature of spiritual experiences.[50] In addition, they may help individuals attain positive behaviours, such as moral insight and empathy, although these could be accompanied by intolerance and fanaticism.[51]

From this information one can conclude that religious experiences are accompanied by changes in neural states. This is an obvious and relatively uninteresting observation, although it does underline a reality the Christian community should not ignore. Second, one would like to know whether some individuals are more amenable than others to these brain changes, and hence whether it is easier for some to experience this particular religious phenomenon than for others. Third, if it is possible to induce these brain changes by psychological or pharmacological means, enormous caution would be required in interpreting the resulting phenomenon as having any religious significance under those circumstances.

---

49   A. Newberg et al., 'The Measurement of Regional Cerebral Blood Flow During the Complex Cognitive Task of Meditation: A Preliminary SPECT Study', *Psychiatry Research: Neuroimaging* 106/2 (2001), 113–22; A. Newberg et al., 'Cerebral Blood Flow During Meditative Prayer: Preliminary Findings and Methodological Issues', *Perceptual and Motor Skills* 97/2 (2003), 625–30.

50   Patrick McNamara, 'The Motivational Origins of Religious Practices', *Zygon* 37 (2002), 143–60.

51   A fascinating exploration of the relationship between neurological illness and visions is provided by Mark Salzman in his novel, *Lying Awake*. A Carmelite nun is regarded as a spiritual master, but is then diagnosed with dangerous headaches, leading to a devastating choice between health and spirituality. Mark Salzman, *Lying Awake* (New York, NY: Alfred A. Knopf, 2000).

While this discussion of the use of neural investigative techniques has contemporary overtones, due to the increasing ability to 'look into' the brain, it is also remarkably reminiscent of long-established psychological procedures for altering people's responses. The well-known realms of mind control or brainwashing are viewed pejoratively on account of their association with unethical uses of manipulative methods to change people's views or attitudes to conform to those of others.[52] Generally psychological in nature, they are forms of behavioural modification, by which suitable psychological pressure using a variety of conditioning techniques can be invoked to entice a person to change their views. While coercion may not always be used (for instance, methods to quit smoking via hypnotism may not be coercive) these methods are characterized by the views of a dominant individual or group being transferred to a submissive individual.[53] The problem with any such approach stems from the lack of well-articulated reasons for the change of perspective. Pressure has been exerted, even unintentionally in some instances, and has been the main factor in the resulting behavioural change. While it may not be known which brain regions have been influenced, the neural effects will be just as definite as direct intrusion into the brain.

When considering religious implications of these forms of behavioural modification, regardless of the procedure employed – direct or indirect, it is important to work from the foundation that all individuals are people made in the image of God, with a resulting dignity and uniqueness. No matter how attractive it may appear to entice someone to adopt one's own perspective (including a Christian one), that change will lack integrity and legitimacy if it is imposed, either psychologically or neurally. It will also represent a flouting of the individual's autonomy and ability to act as a free agent.

52   Dominic Streatfeild, *Brainwash: The Secret History of Mind Control* (London: Hodder & Stoughton, 2006); Kathleen Taylor, *Brainwashing: The Science of Thought Control* (Oxford: Oxford University Press, 2006).

53   J. S. Blumenthal-Barby, 'Between Reason and Coercion: Ethically Permissible Influence in Health Care and Health Policy Contexts', *Kennedy Institute of Ethics Journal* 22/4 (2012), 345–66.

The neuroimaging studies raise questions about the biological basis, function and evolutionary history of religion. However, they cannot address the authenticity of such experiences.[54] This is because one comes up against the ever-present question of which comes first: is it the brain state or the religious experience? Related to this is the possibility that the same brain state may be associated with quite different religious worldviews (as between Roman Catholic and Buddhist). For instance, one would like to know whether one could distinguish between different views on the divinity of Christ from examining people's brains. The precision required here may be forbidding, and yet it may be of far greater relevance than knowing whether someone is 'religious' or 'nonreligious', or even has a tendency towards fundamentalist or liberal perspectives on religious matters. Much of this is speculation, and while it extends well beyond the territory of existing science, it is not too far removed from data that may be accessible in the foreseeable future.

How much insight can neuroscience provide about religion and what will be the nature – let alone value – of that information? Currently investigators are attempting to assess the subjective religious experiences of individuals rather than the shared belief system that is religion embedded in its cultural-historical framework.[55] They are certainly not determining the existence or non-existence of a divine being. But my point remains. We should not be surprised to find that neural correlates exist for fundamental activities within the Christian community.

While these neural issues may seem to occupy a unique situation in the human body, that is far from the case. Take the gastrointestinal tract. People's enjoyment of a good meal is not dependent upon knowing anything about the physiological state of their gastrointestinal tract, how it

---

54   Nina P. Azari, 'Neuroimaging Studies of Religious Experience: A Critical Review', in Patrick McNamara, ed. *Where God and Science Meet: How Brain and Evolutionary Studies Alter Our Understanding of Religion* (Westport, CT: Praeger Publishers, 2006), 33–54.

55   Matthew Ratcliffe, 'Neurotheology: A Science of What?', in Patrick McNamara, ed. *Where God and Science Meet: How Brain and Evolutionary Studies Alter Our Understanding of Religion* (Westport, CT: Praeger Publishers, 2006), 81–104.

functions or what is functioning when the meal is being eaten and enjoyed. These are subjects ripe for scientific investigation, but regardless of the progress made in this regard, a precise scientific description of the processes involved (including the neural processes involved) will or at least should have little, if any, relevance for one's appreciation of food. The quality of a meal, its nutritional status, its tastiness and its presentation are all matters to be judged by their own criteria, and not on the basis of an analysis of the physiological events underlying them. So with an understanding of the brain and religious experiences: each is to be judged on its own merits even though an understanding of neural processes may throw light on certain elements within the religious sphere.

Neuroimaging and moral behaviour

And yet doubts linger. Take the case of the generally agreed distinction between two forms of disgust – visceral and moral. On the surface these appear to be quite different, and yet visceral disgust that is common to human cultures worldwide may have formed the neural basis for the evolutionary development of moral repugnance.[56] Visceral disgust functions to protect bodily purity and integrity, for example by preventing us from eating contaminated food. This 'core' disgust is supposedly associated with socio-moral disgust concerning more abstract issues, such as our reactions to late-term abortion, homosexuality, embryo research or murder.

One fMRI study showed that overlapping brain areas are activated whether individuals experience visceral or moral disgust, the implication being that these emotions are related.[57] Does this mean that there is no

---

56 D. Jones, 'Moral Psychology: The Depths of Disgust', *Nature* 447/7146 (2007), 768–71.

57 J. Moll et al., 'The Moral Affiliations of Disgust: A Functional MRI Study', *Cognitive and Behavioral Neurology* 18/1 (2005), 68–78.

category difference between our responses to contaminated food and late-term abortion? Are our often firmly held moral intuitions therefore, little more than impulsive gut reactions rather than considered moral and/or theological positions?[58] To argue that there are no category differences is a misinterpretation of the fMRI data. The common element is provided by 'disgust', but this tells us nothing about why some people find embryo research, say, disgusting, but others do not. Moral judgements are not implicitly tied in to feelings of disgust, since levels of moral disgust can decrease (or increase) as we ponder the issues at stake.

Along similar lines, the case has been made that donations made to charitable causes activate the 'reward system' in the brain in similar fashion to its activation by food, drugs and sex. In this fMRI study it was found that altruistic acts, such as giving away money, lit up the primitive mesolimbic reward system in the brain.[59] A straightforward interpretation is that altruism ushers in a sense of reward; it feels good to do good. But this should not be extrapolated to conclude that people give away money with the prime or even sole reason that it will make them feel good. The authors, however, go even further and conclude that performing charitable acts may be hard-wired into the brain; according to them, these acts are not a product of culture or, one assumes, of moral reflection. This is over interpretation of the data, in addition to which fMRI images are based on nothing more than changes in blood flow in the brain regions concerned. While these changes are not to be idly dismissed, the conceptual gap between them and conclusions regarding behaviour, altruism in this instance, is debatable.

Similar comments can be made about another study in the same genre. In this instance, fMRI was employed to examine the brains of subjects set the task of choosing whether to voluntarily give money to a food bank, or

---

58   M. Midgely, 'Biotechnology and Monstrosity', *Hastings Center Report* 30 (2000), 7–15.

59   J. Moll et al., 'Human Fronto-Mesolimbic Networks Guide Decisions About Charitable Donation', *Proceedings of the National Academy of Sciences of the United States of America* 103/42 (2006), 15623–8.

through mandatory taxation.[60] Surprisingly, perhaps, even when the money went to the food bank via taxation, the reward centre in their brains lit up. The authors concluded that pure altruism does exist, since satisfaction was derived from an increase in the public good in the absence of the reciprocal benefit, a 'warm glow' that the giver would receive from voluntary giving. However, activation of the brain region was greater when the money was voluntarily given. While these results can be interpreted in different ways, it is worth noting that two of the three authors were economists, whose primary interest was in determining taxation policy rather than in discovering how the brain works. It may be that the results are more enlightening to neuroeconomists than to neuroscientists, let alone theologians.

Regardless of the evidential basis for the conclusions reached, they present a renewed challenge to moral and theological decision-making, namely, to demonstrate that acts of kindness and altruism are indeed motivated by compassion and moral feeling rather than by a primitive urge for the good feeling produced by neural events. Renewed efforts are needed to provide a thoroughly grounded conceptual basis for the validity of altruism; otherwise, it becomes all too easy to assert that it amounts to little more than a drive like that for food or sexual satisfaction. The relationship between the rationale underlying altruistic acts and their neural basis is in urgent need of clarification. While I have no problem in contending that a neural basis for such drives does not in itself undermine our moral faculty, since the neural events are in no way causative, the task of substantiating this will be ongoing. From my perspective this is a task that should be welcomed by theologians as a means of widening our horizons on the contribution that neuroscience can make to theology.

Regardless of such provisos, neuroimaging is being increasingly presented as evidence in courts of law to help determine culpability. In a number of high-profile cases the defence has sought to admit brain images

---

60   W. T. Harbaugh, U. Mayr, and D. R. Burghart, 'Neural Responses to Taxation and Voluntary Giving Reveal Motives for Charitable Donations', *Science* 316/5831 (2007), 1622–5.

as evidence of mitigated responsibility for criminal actions.[61] While this
has immediate consequences for the legal profession, it also has implications for Christian thinking around the notion of moral responsibility.

One of the great problems is that brain images are visually arresting,
and hence may prove dangerously persuasive, giving the impression of
greater certainty than is scientifically justifiable.[62] This apparent certainty
is misleading, masking as it does the social and family context within which
the individual concerned was raised, educated and later lived (see the case
of James Fallon in chapter 4). It also pays little, if any, attention to the
belief system of the individual and the role this may have played in their
actions. Consequently, brain images should only be used in a court of law
to establish a correlation between a structural abnormality and a specific
deficit, not to demonstrate motivation, responsibility or a predisposition
towards a particular behaviour. Conclusions any firmer than this are premature considering our relatively poor understanding of the brain and its
complex interactions. Nevertheless, even a moral evil, such as violence, or
a moral good, such as altruism, has a neural substrate.

All too readily writers can fall into the trap of claiming that religious
sentiments are 'nothing but' a matter of neural organization, or 'nothing
but' the outpouring of certain neurotransmitters. The conclusion: what
are needed are neurotransmitters, not prayer! This gives the appearance
of being a re-run of neural determinism. However, correlations do not
provide immediate answers regarding causation. In addition, is there a
pathological ingredient? Neural pathologies give rise to experiences that for
some have religious overtones, just as other pathologies wipe out previous

61   J. H. Baskin, J. G. Edersheim, and B. H. Price, 'Is a Picture Worth a Thousand Words?
      Neuroimaging in the Courtroom', *American Journal of Law & Medicine* 33/2–3
      (2007), 239–69; L. S. Khoshbin and S. Khoshbin, 'Imaging the Mind, Minding the
      Image: An Historical Introduction to Brain Imaging and the Law', *American Journal
      of Law & Medicine* 33/2–3 (2007), 171–92.
62   L. R. Tancredi and J. D. Brodie, 'The Brain and Behavior: Limitations in the Legal Use
      of Functional Magnetic Resonance Imaging', *American Journal of Law & Medicine*
      33/2–3 (2007), 271–94.

religious commitments.[63] In these instances, the task is to elucidate how the behaviour and belief patterns of the individual before the illness have been modified by the pathological phenomenon. To overlook the role of the abnormality is to fall into the trap of equating pathology with normality.

Technologically enhancing brains

This topic was touched on in chapter 2 and I will return to it in chapter 6 in the context of technologies postulated to enhance morality. Enhancing brains depends upon the ability to modify people's brains through the use of drugs to increase or decrease the levels of neurotransmitters in targeted brain regions. Intrusions of this order could be used for therapeutic or enhancement purposes, or to modify decision-making abilities. As with all technologies, there is ample room for every kind of good and evil use. The question here is to ask to what extent Christians have begun to come to grips with these developments, since they have major pastoral implications as well as fundamental conceptual ones.

A useful illustration is provided by efforts at enhancing performance, such as cognitive enhancement. Discussions generally centre on the use of drugs such as Ritalin (methylphenidate) that stimulates the brain and increases levels of the neurotransmitter dopamine. Its use to help people stay awake and alert for longer is non-prescription (non-therapeutic) use.[64] Another drug of considerable relevance is Modafinil that appears to be implicated in improving concentration, and in increasing alertness and

63   Jeffrey L. Saver and John Rabin, 'The Neural Substrates of Religious Experience', *The Journal of Neuropsychiatry and Clinical Neurosciences* 9 (1997), 498–510.

64   B. Sahakian and S. Morein-Zamir, 'Professor's Little Helper', *Nature* 450 (2007), 1157–9.

wakefulness.[65] Then there is Donepezil (Aricept) that may be taken to bolster recall of training in pilots.[66]

Propranolol, a beta-blocker, but with a reputation for blocking the formation of traumatic memories,[67] represents a further foray into this fascinatingly tantalizing territory.[68] Non-conscious pathological memories can arise from trauma such as in combat, rape and horrific natural disasters, and may result in post-traumatic stress disorder (PTSD). By administering beta-blockers such as propranolol just before or after the traumatic event it is possible to prevent the embedding of pathological memories of fearful events.[69] Alternatively, if administered during flashbacks some time after the event it is possible to erase the pathological memories.[70] However, such drugs could also be taken to erase the kind of unpleasant, but non-pathological, memories that are generally considered integral to normal human life, such as in response to a near-fatal car accident, the loss of a loved one, or the suffering associated with severe pain from an acute medical emergency. A possible extension of this could see the development of drugs to

65   Michael J. Minzenberg and Cameron S. Carter, 'Modafinil: A Review of Neurochemical Actions and Effects on Cognition', *Neuropsychopharmacology* 33/7 (2007), 1477–502; Brian Vastag, 'Poised to Challenge Need for Sleep, "Wakefulness Enhancer" Rouses Concerns', *JAMA: The Journal of the American Medical Association* 291/2 (2004), 167–70.

66   Georg Grön et al., 'Cholinergic Enhancement of Episodic Memory in Healthy Young Adults', *Psychopharmacology* 182/1 (2005), 170–9; Gary Lynch and Christine M. Gall, 'Ampakines and the Threefold Path to Cognitive Enhancement', *Trends in Neurosciences* 29/10 (2006), 554–62.

67   R. K. Pitman et al., 'Pilot Study of Secondary Prevention of Posttraumatic Stress Disorder with Propranolol', *Biological Psychiatry* 51/2 (2002), 189–92.

68   Walter Glannon, 'Psychopharmacology and Memory', *Journal of Medical Ethics* 32 (2006), 74–8.

69   Pitman et al., 'Pilot Study of Secondary Prevention of Posttraumatic Stress Disorder with Propranolol'; G. Vaiva et al., 'Immediate Treatment with Propranolol Decreases Posttraumatic Stress Disorder Two Months after Trauma', *Biological Psychiatry* 54/9 (2003), 947–9.

70   A. Brunet et al., 'Effect of Post-Retrieval Propranolol on Psychophysiologic Responding During Subsequent Script-Driven Traumatic Imagery in Post-Traumatic Stress Disorder', *Journal of Psychiatric Research* 42/6 (2007), 503–6.

remove all traces of guilt, shame or grief in healthy individuals.[71] This is speculative, and such far-reaching effects may never eventuate. Were they to do so, the theological ramifications would be major, since they would intrude into the inner sanctum of human existence, impinging upon a central aspect of our humanity.

These secondary uses of psychoactive drugs take them out of the traditional realm of therapy into a realm that is 'beyond therapy'.[72] The question is: does this matter, either ethically or theologically? Why not make people 'better than well' by transforming a shy person into a vivacious one, a risk-averse person into a risk-taker, an irresponsible individual into a deeply religious one, simply by the application of increasingly sophisticated pharmaceutical agents? Perhaps there is a moral obligation to exploit these, as long as the people concerned are not harmed. But would this be tampering with the essence of what it means to be human, and what does that mean?

It is very easy to get carried away by such far-reaching possibilities, and overlook the simple fact that these drugs are affecting the most complex of organs, the brain. An underlying assumption is that these drugs are effective and are also free of side effects. In other words, the picture so frequently painted is that they do everything one could possibly hope for and without any negative impact. This is not the case on either score. The reality is that the most promising drugs currently used for cognitive enhancement are addictive. For instance, the mechanisms in the brain for learning and memory are closely connected with addictive behaviour; hence issues connected with the use of Modafinil.[73] Consequently, there is a major distinction between technological innovations such as cell phones or computers, or even cups of coffee, and the use of drugs that intervene directly in the neurobiological basis of one's personality.[74]

71   Glannon, 'Psychopharmacology and Memory'.
72   President's Council on Bioethics, *Beyond Therapy: Biotechnology and the Pursuit of Happiness* (Washington, DC, 2003).
73   Heinz et al., 'Cognitive Neuroenhancement'.
74   Heinz et al., 'Cognitive Neuroenhancement'; Andreas Heinz et al., 'Identifying the Neural Circuitry of Alcohol Craving and Relapse Vulnerability', *Addiction Biology* 14/1 (2009), 108–18.

The difference lies in the externality or transitory nature of the former, as opposed to the more profound and longer lasting effects of the neuroenhancers.

All these are examples of cognitive enhancement such as reasoning, perception, memory, and judgement. It is the augmenting of some aspect of human intellect, providing people with better comprehension of complex situations, or enabling them to devise speedier and better solutions to problems. However, the dividing line between normality and abnormality, and between therapy and enhancement, is fraught. It has become difficult in a number of situations to decide whether one is dealing with genuine medical treatment or social manipulation. It all depends on one's perspective and expectations, and also on the extent to which one is prepared to attempt to solve complex behavioural challenges by direct modification of the brain.

What might be the core of a Christian response to cognitive enhancements? Their mundane nature is their allure, but also their deceptiveness. Some argue that all enhancement is to be eschewed in favour of acceptance of the 'given'.[75] However, considering Christianity's characteristic embrace of the healing ministries and the blurred distinction between therapy and enhancement, this stance is difficult to defend on theological grounds. For instance, theologian and ethicist, Ted Peters questions whether a Christian faith that emphasizes redemption should not also embrace 'all forms of human betterment, even enhancement'.[76] For him a holistic view of health as frequently championed by Christian anthropology may even have space for the enhancement of the social and relational aspects of our humanity. These pointers are at odds with the precautionary stance often encountered in Christian thinking that tends towards acceptance of the status quo and expresses doubt about technological interference.

75   M. J. Sandel, *The Case against Perfection: Ethics in the Age of Genetic Engineering* (Cambridge, MA: Belknap Press, 2007).
76   Ted Peters, 'The Soul of Trans-Humanism', *Dialog: A Journal of Theology* 44/4 (2005), 381–95.

Savulescu and Sandberg have taken the neuroenhancement debate even further by proposing the use of psychopharmaceuticals to enhance romantic love and marriage.[77] They suggest that artificially manipulating levels of testosterone, oxytocin and other hormones may help decrease the rate of divorce by enhancing pair-bonding and attachment. We may or may not take this suggestion seriously, but it does force us to ask whether there are morally relevant differences between counselling and neurostimulation. In my view there are, since the one taps into human responsibility whereas the other completely bypasses it. Ready acceptance of the latter appears to reduce human beings to psychological machines, controlled by hormonal and neurotransmitter levels and nothing more. It is the 'nothing more' that is the crucial marker of a deterministic world of psychological impulses and responses.

Proposals along these lines constantly prompt us to revisit the distinction between the brain and the person, the physical biological unit and the whole being that transforms that biological unit into a living being, the material organization of the body and brain and the responsible 'I'. This is why the earlier section on 'brains and person' is critical when faced by technologically grounded scenarios that feed on our imagination. Is it 'I' who is responding to a loved one or contributing to a good cause, or is it 'my brain' that is responding in a way determined by a cocktail of drugs or electrical stimuli applied to my brain? The latter is one of the concerns of some patients with Parkinson's disease who have electrodes implanted to control their aberrant motor movements. There is an artificiality to being able to switch something on and off at will, a phenomenon that sets even useful technological intrusions apart from that which we normally experience. The following section will take this further.

77   J. Savulescu and A. Sandberg, 'Neuroenhancement of Love and Marriage: The Chemicals between Us', *Neuroethics* 1 (2008), 31–44.

Deep brain stimulation and brain machine interfaces

Deep brain stimulation (DBS) is used in patients with Parkinson's disease, in whom electrical signals generated in a subcutaneously placed unit are sent to electrodes implanted in the motor regions of the brain. The aim of these is to stimulate the function of the motor regions that have been detrimentally affected by the loss of the neurons producing the neuro-transmitter dopamine, in an attempt to control motor activities.[78] It is used when routine treatments have become ineffective, although there may be negative side effects including personality changes.[79] Worldwide more than 80,000 patients have been provided with these implants. DBS is also used as an experimental treatment for intractable depression and obsessive-compulsive disorder.[80] While not all patients respond to the treatment, in many the primary symptoms are substantially improved with rare adverse effects.

A range of post-operative neuropsychiatric symptoms has been reported when DBS is used for Parkinson's disease, including depression and apathy, though most are transient and treatable.[81] If side effects of this nature are minor, the alleviation of the crippling motor deficiencies will be welcomed. The balance between the positives and negatives will weigh strongly in the positive direction, and will be assessed as clinically accept-able. The underlying assumption is that there are no noticeable effects on the patients' identity. The change is strictly therapeutic, and equates with any other form of therapy to alleviate the troubling symptoms. Overall,

---

78 D. Gareth Jones and Maja I. Whitaker, 'Transforming the Human Body', in C. Blake, C. Molloy, and S. Shakespeare, eds, *Beyond Human: From Animality to Transhumanism* (London: Continuum, 2012), 254–79.

79 Walter Glannon, 'Stimulating Brains, Altering Minds', *Journal of Medical Ethics* 35/5 (2009), 289–392.

80 J. Kuhn et al., 'Deep Brain Stimulation for Psychiatric Disorders', *Deutsches Ärzteblatt International* 107/7 (2010), 105–13.

81 V. Voon et al., 'Deep Brain Stimulation: Neuropsychological and Neuropsychiatric Issues', *Movement Disorders* 21/Suppl 14 (2006), S305–27.

DBS is an example of a relatively successful neural prosthesis, and illustrates a melding of brain and machine.

A related development is that of neuroprostheses that allow patients with these brain implants to manipulate a cursor on a screen, by directing their thoughts to a specific action.[82] This can then allow the individual to interact with a computer, and can for instance send emails and control a television. When used by 'locked-in' patients this could allow them to communicate with the world in an unprecedented manner.[83]

Neuromotor prostheses have also been used to control the movement of a robotic device.[84] Motor signals are recorded through electrodes implanted in the brain and sent to a unit that computes and generates signals to direct an artificial limb.[85] Here there are two levels of artificiality – the limb and the interface in the brain. Along similar lines one can think of spinal implants to help paraplegics walk.[86] The risks associated with such an invasive approach and the long-term care required for this type of implant are considerable.[87]

Examples such as these may not elicit undue concern, since they are aiming to replace functions and abilities that have been lost. In regaining

82   L. R. Hochberg et al., 'Neuronal Ensemble Control of Prosthetic Devices by a Human with Tetraplegia', *Nature* 442/7099 (2006), 164–71; S. I. Ryu and K. V. Shenoy, 'Human Cortical Prostheses: Lost in Translation?', *Neurosurgical Focus* 27/1 (2009), E5.

83   N. Birbaumer, 'Breaking the Silence: Brain-Computer Interfaces (BCI) for Communication and Motor Control', *Psychophysiology* 43/6 (2006), 517–32; A. Fenton and S. Alpert, 'Extending Our View on Using BCIs for Locked-in Syndrome', *Neuroethics* 1 (2008), 119–32.

84   Hochberg et al., 'Neuronal Ensemble Control of Prosthetic Devices by a Human with Tetraplegia'.

85   E. A. Pohlmeyer et al., 'Toward the Restoration of Hand Use to a Paralyzed Monkey: Brain-Controlled Functional Electrical Stimulation of Forearm Muscles', *PLoS One* 4/6 (2009), e5924; M. Velliste et al., 'Cortical Control of a Prosthetic Arm for Self-Feeding', *Nature* 453/7198 (2008), 1098–101.

86   A. J. Fong et al., 'Recovery of Control of Posture and Locomotion after a Spinal Cord Injury: Solutions Staring Us in the Face', *Progress in Brain Research* 175 (2009), 393–418.

87   Ryu and Shenoy, 'Human Cortical Prosthesess'.

lost function the patient is liberated and normalized. It is possible that the mechanical nature of the intervention may have an alienating effect, although even this may be temporary and the disadvantages may pale into insignificance compared with the functional advantages. Adapting to the newness and unusual nature of the artificial devices involved may be no more demanding than learning to live with an artificial limb or renal dialysis.

The examples just outlined are examples of brain machine interfaces (BMIs), and fall within the bounds of human experience. We are capable of envisaging them and thinking around their implications. Ethically, we are aware of the centrality of fully informed consent, especially where the processes of interest are experimental in nature. They pose no threat to our humanness, and encouragingly show the promise of medical technology, a promise that can all-too-readily become submerged beneath concerns at the pitfalls. But as technology is increasingly internalized does this threaten the integrity of the human body? As with a number of other areas I have touched on in other chapters, debate on BMIs tends on occasion to be unduly expansive. Consider the following quote: 'There is both a sense of human fulfillment and of human transgression when BMIs are imagined, both a sense that BMI is a form of transcendence that will allow us to realize heretofore unknown aspects of our humanity and a sense that eventually BMI will eclipse humanity altogether.'[88]

This quotation introduces the realm of hypothesis and speculation that may be far from profitable and raises a host of demanding and frustrating questions. How would the interaction between mind and machine impact on personhood? Does mental enhancement by BMIs alter the personality of the user? Might there be unwanted consequences for the individual's personality? This is not the case at present, but with the development of increasingly sophisticated and intrusive interfaces this may change. It may be that one day there will be direct interaction between internalized computers in our brains and neural processes and decision-making. The involvement of machine interfaces could perhaps alter an individual to

---

88   Michael L. Spezio, 'Brain and Machine: Minding the Transhuman Future', *Dialog: A Journal of Theology* 44/4 (2005), 375–80.

such a degree as to threaten the persistence of personal identity.[89] This may have flow-on effects for the attribution of moral responsibility. These questions and possibilities, speculative though they are, have a threatening air to them, and so we need to return to basics.

If the reason for moving in the direction of BMIs is to overcome a pathological deficit, their use may well prove beneficial, a la conventional therapies. They may increase autonomy by restoring a lost capability. On the other hand, if used for insubstantial reasons the opposite may be the end result: jeopardizing autonomy and serving as a threat to wellbeing. These are hardly unique ethical considerations. The goal is to benefit patients and enable them to live fuller lives as the people they and others know them to be. It is to enhance their standing as people made in the image of God.

Concluding comments

This chapter has traversed considerable territory, and it may seem that technological intrusions into the brain, such as DBS, are far removed from the more erudite concerns of dualism and personhood. However, the aim has been to show that the latter debate is highly relevant for an understanding of how injury and disease affect brain function and can be ameliorated using a range of technological tools. It is only as the nature of the physical brain is appreciated, that it becomes possible to assess the impact of therapeutic measures on the individual as a whole person. This in turn opens the way to assessing the value of these measures within a Christian context. It also begins to reflect some of the problems with a view of the human individual as a dualistic entity, with brain and soul/mind as essentially distinct compartments. The person-centered physicalist

---

89   Guglielmo Tamburrini, 'Brain to Computer Communication: Ethical Perspectives on Interaction Models', *Neuroethics* 2 (2009), 137–49.

position adopted here provides a framework within which injury and disease, alongside learning and adaptation, can be accommodated in a system open to both human and divine influence. It also provides a basis for critiquing excessive technological intrusions that could threaten the wellbeing and freedom of humans individually and in community.

CHAPTER 6

Technological enhancement of morality

An emerging theme within an increasing amount of current thinking centres on ways in which humans can transcend their humanness, or become more than human. Looy alludes to this in her survey of the theological frontiers of psychology.[1] This is one of the most provocative frontiers promulgated by those who wish to transform the dimensions of human nature. While there are many facets to this endeavour, the one that is both best known and also most extreme is that of transhumanism, with its myriad goals of not only dramatically extending human abilities and life span technologically, but also finding ways of overcoming the burden of our mortality.[2] However, there are many who may not consider themselves transhumanists, but who have vast agendas for enhancing human cognitive abilities, and even some who see it as their task of advocating for the enhancement of specifically moral attitudes.[3] The thrust in all cases is to accomplish these ends utilizing the latest developments in biomedical technology. It is this that sets them apart from so much that has gone on in the past, and that is based on the alleged infinitely malleable nature of the human body and brain.[4]

1   Heather Looy, 'Psychology at the Theological Frontiers', *Perspectives on Science and Christian Faith* 65/3 (2013), 147–55.

2   J. Garreau, *Radical Evolution: The Promise and Peril of Enhancing Our Minds, Our Bodies – and What It Means to Be Human* (New York, NY: Doubleday, 2005).

3   Ingmar Persson and Julian Savulescu, *Unfit for the Future: The Need for Moral Enhancement* (Oxford: Oxford University Press, 2012).

4   The essence of this chapter has appeared as an article in a special issue of *Perspectives on Science and Christian Faith*, on Psychology at the Theological Frontiers. D. Gareth Jones, 'Moral Enhancement as a Technological Imperative', *Perspectives on Science and Christian Faith* 65/3 (2013), 187–95.

Hence two characteristics immediately become evident: human beings are ultimately in control and are the final arbiters of all they are, and control stems from the power wielded by biomedical technology. By implication there is no place for God, and no hope exists beyond that which is at the behest of human beings and can be directed by human technology.

These characteristics constitute the fundamental underpinning of this whole endeavour although rarely are they made explicit. If God does not exist humans are left to work out their morality for themselves, and if found wanting, it follows that they have recourse to whatever procedures are at their disposal. This is where the present debate on using technological means of enhancing moral behaviour becomes relevant. If unenhanced moral attitudes prove inadequate, why not enhance them in directions that will improve the quality of life of individuals as well as of society?

The possibilities opened up by some form of technologically assisted moral enhancement immediately raise questions. Will the technology work? If it does, can it be utilized without contravening basic ethical values, such as the autonomy and dignity of those being enhanced? If these provisos can be satisfactorily addressed, who is to decide what enhanced moral behaviour is to look like? Will the decision-makers themselves be enhanced or will they simply be those with power and status?

While the technology currently being contemplated is new, the prospects opened up by improving and/or controlling morality are far from new. It has been the domain of science fiction writers for many years, and as is so often the case, their prescience is striking.

Morality in a bottle

And there's always soma to calm your anger, to reconcile you to your enemies, to make you patient and longsuffering. In the past you could accomplish these things by making a great effort and after years of hard moral training. Now, you swallow two or three half-gramme tablets, and there you are. Anybody can be virtuous now.

You can carry at least half your morality about in a bottle. Christianity without tears – that's what soma is.[5]

So said Mustapha Mond, the Controller, in *Brave New World*, written eighty years ago. Aldous Huxley was looking into the future, but as with so many of the other possibilities he raises, the tantalizing images bring us face to face with a world that is far closer to ours than we are frequently prepared to accept. The brave new world he envisaged was far from a paradise; why should our brave new world be any different? And yet many find it difficult to accept that biomedical technology will be unable to give us all we want or could ever desire. The warnings contained in *Brave New World* are ignored, perhaps because the context of the world envisaged by Huxley differs considerably from that pictured by contemporary bioethicists. Consider the following comments made by bioethicists and medical technologists in recent years.

The high profile bioethicist, Peter Singer, has proposed a 'morality pill'. Since moral behaviour is in part biochemically determined, it should be possible he argues to engineer moral behaviour with drugs. Imagine the following: 'If continuing brain research does in fact show biochemical differences between the brains of those who help others and the brains of those who do not, this pill should be taken by those who do not normally help others'.[6]

Julian Savulescu, another well-known bioethicist, has argued that, 'If safe moral enhancements are ever developed, there are strong reasons to believe that their use should be obligatory, like education, or fluoride in the water, since those who should take them are least likely to be inclined to use them. That is, safe, effective moral enhancement would be compulsory'.[7]

---

5   Aldous Huxley, *Brave New World*, 1958 ed. (Harmondsworth: Penguin Books, 1932), 185.
6   Peter Singer and Agata Sagan, 'Are We Ready for a "Morality Pill"?', *New York Times* (28 January 2012) <http://opinionator.blogs.nytimes.com/2012/01/28/are-we-ready-for-a-morality-pill/> accessed 23 June 2013.
7   Ingmar Persson and Julian Savulescu, 'The Perils of Cognitive Enhancement and the Urgent Imperative to Enhance the Moral Character of Humanity', *Journal of Applied Philosophy* 25/3 (2008), 162–77.

The bottle in modern parlance may deviate from what Huxley had in mind, but the thrust is remarkably similar – a pharmaceutical product of some description will transform morality. Something externally administered will make a substantial difference to the individual by changing or even transforming attitudes and aspirations, and even perhaps abilities and predispositions. In practice this means modifying neural processes within the brain as discussed in chapter 5.

The enhancement literature is plagued by confusion about the definition of the term and also about its delineation from therapy. In part the confusion stems from different conceptions of what constitutes enhancement, the areas of overlap between therapy and enhancement (regardless of definitions), and the extent to which the one blends into the other. My stance is that there is a continuum from unambiguous therapy (removing an appendix that is about to rupture) at the one end, to unambiguous enhancement (curing death and creating posthumans to live for a few hundred years) at the other. In between one can think of the enhancement of healthy people by the use of vaccines as prophylactics.[8]

As emerged in chapter 5, it would be a major mistake to think that this discussion is of some far-distant future, only of interest to science fiction writers and futurologists. The possibility of enhancing the lives of ordinary people through biomedical technology is present reality. Examples abound as drugs originally designed to treat a medical condition are employed by healthy individuals to improve their performance. For instance, up to 25 per cent of American students use psychostimulants,[9] while 5 per cent of the working population in Germany are reported as using pharmaceutical drugs to enhance their cognitive functions. It has also been claimed that up to 80 per cent of students in Germany would use neuroenhancers if they could be assured there would be no adverse effects.[10] In chapter 5 we encountered examples of the secondary uses of drugs that had origi-

---

8   Jones, 'Enhancement: Are Ethicists Excessively Influenced by Baseless Speculations?'.
9   Henry Greely et al., 'Towards Responsible Use of Cognitive-Enhancing Drugs by the Healthy', *Nature* 456 (2008), 702–5.
10   See Heinz et al., 'Cognitive Neuroenhancement'.

nally been designed for therapeutic purposes, but are now being employed by healthy individuals to stave off tiredness, improve concentration and short-term memory and combat the formation of traumatic memories.

While such examples of cognitive enhancement are current reality, the intention of more sophisticated approaches is to enable individuals to demonstrate improved moral behaviour. Here the intention is to make individuals more self-sacrificial, empathic and altruistic, or to decrease their impulses towards violence and aggression. Perhaps they (and we) could be made more loving and even more spiritual. On the surface it is preferable to have a cooperative and intelligent employee, an honest researcher, and a compliant student or prisoner rather than the converse.

There seems little doubt that biochemical interactions stimulate our moral imagination, increase our empathy towards others, or, in the cognitive dimension, improve our powers of moral judgement and reasoning. It has been argued that even memory-improving drugs may contribute to moral improvement by enabling the fallible memory recall the truth, thereby helping to avoid self-deception.[11] Of course, the converse may also apply.

An essential proviso would appear to be that these methods for improving moral decision-making actually do improve it, and that they are more effective than conventional approaches. Any claims that they are more effective should be open to scientific scrutiny, since what is being conducted is a scientific experiment. This should apply to any new treatment, and there is no reason why moral enhancement procedures are excluded from stringent analysis and critique. In clinical practice we do not accept the validity of new treatments based on the positive reports of patients or the unsubstantiated claims of clinicians. Publication of results, peer review of the publications, and openness to testing and retesting are seen as basic requirements. Why then should claims regarding the effectiveness and desirability of 'moral technology' be any different?

11   Allen Buchanan, *Better Than Human: The Promise and Perils of Enhancing Ourselves* (Oxford: Oxford University Press, 2011).

The mechanics of moral bioenhancement

The scientific basis for thinking about moral bioenhancement encapsulates a variety of approaches. The first of these is transcranial direct current stimulation (TDCS). It has recently emerged that TDCS can be used to improve language and mathematical abilities, memory, problem solving, attention and even movement. In TDCS, weak electrical currents are applied for about twenty minutes to the head via electrodes placed on the scalp. The currents pass through the skull and alter spontaneous neural activity. They are thought to increase neuroplasticity, making it easier for neurons to fire and form the connections that enable learning.[12] It is thought that the effects of TDCS can persist for up to twelve months.[13]

Experiments in humans have found that following TDCS there are changes in the local concentration of the neurotransmitters GABA and glutamate, both of which are important in synaptic mechanisms implementing learning and memory.[14] These characteristics of TDCS make it an attractive tool for manipulating neurobehavioural plasticity and it may be useful in enhancing psychological functions.[15]

Like all technologies TDCS will probably come with costs as well as benefits. Enhancing some capacities may lead to deterioration of others. What this means is that highly developed capacities in some cognitive

---

12   Roi Cohen Kadosh et al., 'The Neuroethics of Non-Invasive Brain Stimulation', *Current Biology* 22/4 (2012), R108–11.

13   C. A. Dockery et al., 'Enhancement of Planning Ability by Transcranial Direct Current Stimulation', *The Journal of Neuroscience* 29/22 (2009), 7271–7.

14   Charlotte J. Stagg and Michael A. Nitsche, 'Physiological Basis of Transcranial Direct Current Stimulation', *The Neuroscientist* 17/1 (2011), 37–53.

15   Michael A. Nitsche et al., 'Transcranial Direct Current Stimulation: State of the Art 2008', *Brain stimulation* 1/3 (2008), 206–23; K. S. Utz et al., 'Electrified Minds: Transcranial Direct Current Stimulation (tDCS) and Galvanic Vestibular Stimulation (GVS) as Methods of Non-Invasive Brain Stimulation in Neuropsychology – a Review of Current Data and Future Implications', *Neuropsychologia* 48/10 (2010), 2789–810.

domains may be accompanied by reduced functioning in others.[16] The success of TDCS in apparently improving some mathematical abilities and the low cost of the equipment is leading to pressure to use it in the home setting, leading to ethical and safety issues, and calls for discussions around the need for a regulatory framework.[17]

While TDCS is a form of cognitive enhancement, some use it as a launch pad into the moral realm. This is, of course, speculative but some argue that certain biochemical interactions 'might stimulate our moral imagination, increase our empathy towards others, [...] improve our powers of moral judgment and reasoning'.[18] What one detects here is a tendency commonly encountered in the bioethical literature, and this is that tentative data are viewed in an unreservedly positive light. The deficiencies and possible drawbacks to a procedure are downplayed in favour of what are seen as its positive aspects, no matter how tentative some of these may be. It is salutary to note this since bias in favour of a positive outlook reflects an unnervingly hubristic mentality.

Better known is the potential contribution of neurotransmitters and neuropeptides. There appear to be brain circuits active during moral judgement that are linked to pro-social emotions such as empathy, guilt and pity.[19] In connection with this it is not unusual to encounter papers

---

16   Kadosh et al., 'The Neuroethics of Non-Invasive Brain Stimulation'.

17   Ewen Callaway, 'Shocks to the Brain Improve Mathematical Abilities', (16 May 2013) <http://www.nature.com/news/shocks-to-the-brain-improve-mathematical-abilities-1.13012> accessed 23 June 2013; Nicholas S. Fitz and Peter B. Reiner, 'The Challenge of Crafting Policy for Do-It-Yourself Brain Stimulation', *Journal of Medical Ethics* (2013), Advance online publication, 3 June 2013, doi: 10.1136/medethics-2013-101458; D. Fox, 'Brain Buzz', *Nature* 472/7342 (2011), 156–8.

18   Allen Buchanan, 2012, quoted in R. Andersen, 'Why Cognitive Enhancement Is in Your Future (and Your Past)', *The Atlantic* (6 February 2012) <http://www.theatlantic.com/technology/archive/2012/02/why-cognitive-enhancement-is-in-your-future-and-your-past/252566/> accessed 23 June 2013.

19   R. J. Blair, 'The Amygdala and Ventromedial Pre-Frontal Cortex in Morality and Psychopathy', *Trends in Cognitive Science* 11 (2007), 387–92; J. Moll et al., 'Functional Networks in Emotional and Nonmoral Social Judgements', *NeuroImage* 26 (2002), 696–703.

with titles such as: 'Serotonin selectively influences moral judgment and behaviour through effects on harm aversion' and 'Oxytocin increases trust in humans'.[20] Both these direct our attention to the two compounds on which most attention is paid in the brain-behaviour relationship: serotonin and oxytocin. They also point to a well-known and well-accepted dictum, namely, that we cannot begin to understand human behaviour without some understanding of the brain.

Serotonin is being put forward as the neural substrate of ethical decision-making.[21] There is evidence that serotonin selectively influences moral judgement and behaviour through increasing subjects' aversion to personally harming others. Administration of a serotonin reuptake inhibitor (SSRI) modulates decision-making in moral dilemmas. Consequently, enhancing serotonin makes subjects more likely to consider that harmful actions should be forbidden. Enhancing serotonin levels changes decision making in a test known as the 'ultimatum game', in that it makes subjects less likely to reject unfair offers. Additionally, this has a stronger effect on people who self-identify as being more empathic.[22]

This is one side of the story regarding serotonin, but there is another and this is that low serotonin levels are associated with self-harm in those who are depressed and inclined towards suicide. Those studying morality do so on healthy subjects, whereas patients with dysfunctional attitudes point to a different facet of serotonin's effects on behaviour. For the latter patients disruption of the serotonin system is consistently associated with non-suicidal self-injury, and suicide in adults and low levels may explain pessimistic dysfunctional attitudes associated with major

20   M. J. Crockett et al., 'Serotonin Selectively Influences Moral Judgment and Behavior through Effects on Harm Aversion', *Proceedings of the National Academy of Sciences* 107/40 (2010), 17433–8; M. Kosfeld et al., 'Oxytocin Increases Trust in Humans', *Nature* 435/7042 (2005), 673–6.
21   Crockett et al., 'Serotonin Selectively Influences Moral Judgment'.
22   See Crockett et al., 'Serotonin Selectively Influences Moral Judgment'; Heike Tost and Andreas Meyer-Lindenberg, 'I Fear for You: A Role for Serotonin in Moral Behavior', *Proceedings of the National Academy of Sciences* 107/40 (2010), 17071–2.

depression.[23] However, there is a complex interrelationship among biological, psychological and social systems.[24]

There seems little doubt that serotonin is influential in human social behaviour, both in health and illness. Consequently one has to be exceedingly careful in thinking that it can be used with impunity to alter moral decision-making in healthy individuals. It is important to ensure that any social dysfunction is principally the result of neural characteristics. Contributions from dysfunctions originating in the environment and in the network of relationships of which the individual is a part should never be peremptorily dismissed.

While the serotonin story is a powerful one, it is impossible to divide the brain into distinct functional compartments. Augmentation of serotonin not only affects behaviour, it is also involved in cardiovascular regulation, respiration, sleep-wake cycles, appetite, pain sensitivity and reward learning.[25] Even within the morality area itself, the enhancement of moral cognition may be accompanied by an increased willingness to allow cheaters to go unpunished. Not only this, in mice there is evidence that enhancing aspects of memory also results in unwanted effects, like higher sensitivity to pain. In other words, improving moral behaviour using a pill is only one means of influencing behaviour. This is because human behaviour, no matter how neurally-based it is, cannot be reduced to simplistic formulae without losing the essence of what it means to be human.

Very similar comments apply to the role of oxytocin, a neuropeptide, in moral enhancement. Once again, the literature is highly dependent upon the results of role-play studies. For instance, the administration of an oxytocin nasal spray increases trust, in that subjects playing the role of an investor appear to be more generous in their investment to a trustee.

23   J. H. Meyer et al., 'Dysfunctional Attitudes and 5-HT2 Receptors During Depression and Self-Harm', *American Journal of Psychiatry* 160/1 (2003), 90–9.

24   S. E. Crowell et al., 'Parent–Child Interactions, Peripheral Serotonin, and Self-Inflicted Injury in Adolescents', *Journal of Consulting and Clinical Psychology* 76/1 (2008), 15–21.

25   Neil Levy, 'Ecological Engineering: Reshaping Our Environments to Achieve Our Goals', *Philosophy & Technology* 25/4 (2012), 589–604.

However, it does not appear to affect an individual's willingness to bear risks in general.[26] In another series of studies it was concluded that oxytocin creates intergroup bias since it motivates in-group favouritism, an important ingredient in cooperation within groups.[27] This suggests it has a role in the emergence of intergroup conflict and violence. While the relationship between oxytocin and trust has created enormous interest, it is important not to treat the data uncritically.

These chemicals, therefore, have a part to play in moral behaviour, and yet one has to question how far this takes us. That they influence some aspects of our moral responsiveness is difficult to reject, but whether this is of fundamental interest morally is open to debate. And in terms of the central task in this chapter, can they be used to morally enhance those in need of moral uplift?

Why the perceived need for moral enhancement?

Over recent years a debate has been raging in the bioethics literature between various prominent bioethicists. This revolves around the following proposition by Persson and Savulescu:

> We claim that human beings now have at their disposal means of wiping out life on Earth and that traditional methods of moral education are probably insufficient to achieve the moral enhancement required to ensure that this will not happen. Hence, we argue, moral bioenhancement should be sought and applied [...] it is a matter of such urgency to improve humanity morally to the point that it can responsibly handle the powerful resources of modern technology that we should seek *whatever* means there are to effect this.[28]

26 Kosfeld et al., 'Oxytocin Increases Trust in Humans'.
27 C. K. W. De Dreu et al., 'Oxytocin Promotes Human Ethnocentrism', *Proceedings of the National Academy of Sciences* 108/4 (2011), 1262–6.
28 Ingmar Persson and Julian Savulescu, 'Getting Moral Enhancement Right: The Desirability of Moral Bioenhancement', *Bioethics* 27/3 (2013), 124–31.

What we have here is a mixture of despair at the plight of the world brought about through the possibilities opened up by scientific and technological prowess, and at the limitations of traditional moral education and discernment. But the irony is that in order to rectify the latter, they look again to technology, this time in the guise of moral bioenhancement.

For Persson and Savulescu, further developments in cognitive enhancement will only make matters worse, since a few people or groups of people will abuse the powers made available to them. Consequently, the priority is to find a way out of the current morass, and for them this is via genetic and other biological means of improving morality. Not only this but, as they argue in other places, this enhancement should be perfected and then made mandatory.[29]

This gets to the core of some of the problematic aspects of the debate: the potential perfectibility of moral enhancement technologies. The likelihood of achieving perfectibility is close to zero. The complexity of the brain is such that it is well nigh impossible to restrict interventions to just one emotion, let alone one moral response. To think otherwise is neuroscientifically naïve. In making a similar point John Harris writes:

> The only reliable methods of moral enhancement, either now or for the foreseeable future, are either those that have been in human and animal use for millennia, namely socialization, education and parental supervision or those high tech methods that are general in their application. By that is meant those forms of cognitive enhancement that operate across a wide range of cognitive abilities and do not target specifically 'ethical' capacities.[30]

And then there is the question of personal liberty; to modulate one's moral responses, if it could be done, would necessitate the imposition of the beliefs and mores of others. Whence freedom, even if the intention was to overcome what are generally regarded as moral evils? And what becomes

29   Persson and Savulescu, 'Getting Moral Enhancement Right'; J. Savulescu, 'Genetic Interventions and the Ethics of Enhancement of Human Beings', in Bonnie Steinbock, ed. *The Oxford Handbook of Bioethics* (Oxford: Oxford University Press, 2007), 516–35.
30   John Harris, 'Moral Enhancement and Freedom', *Bioethics* 25/2 (2011), 102–11.

of Christianity? If freedom of choice has disappeared there is no freedom at all, a deeply disconcerting prospect for Christians, but also for a liberal society. The fundamental guiding principles of contemporary bioethics, namely, autonomy and beneficence, let alone justice, look as though they would have been sacrificed to a technological imperative.

The intentions of writers like Julian Savulescu and Tom Douglas are, for example, to elevate people's responses to the plight of the global poor, or to decrease the harm being caused by a serial philanderer.[31] With these I have much sympathy, and yet the means employed, that of some form of direct emotional modulation, is disconcerting. The second of these examples is probably dysfunctional behaviour, and has to be treated as such. The first is quite different, since it illustrates a lack of empathy with the poor and disadvantaged. Altering emotions, such as sympathy, psychologically or even biologically may leave one's level of practical commitment untouched. That requires moral decision-making based on altruism and siding with the victim. It is a desire to live the good life, and in Christian terms to live for one's neighbour, for the deprived and downtrodden, and for those unable to help themselves. There is a rational basis to moral responsibility, one that involves the whole person and many interrelated regions of the brain.

For Persson and Savulescu there is 'a widening gap between what we are practically able to do, thanks to modern technology, and what we are morally capable of doing, though we might be somewhat more capable than our ancestors were'.[32] For them the drive behind moral bioenhancement is improvement in the powers of reason, impelled by the moral dispositions of altruism and a sense of justice, dispositions that these writers claim have biological bases in evolution. They accept that 'moral bioenhancement worthy of the name is practically impossible at present and might remain so for so long that we will not master it'.[33] They also accept that traditional means of improving moral wisdom are also necessary. Their realism is to

---

31   Thomas Douglas, 'Moral Enhancement Via Direct Emotion Modulation: A Reply to John Harris', *Bioethics* 27/3 (2013), 1601–68.

32   Persson and Savulescu, *Unfit for the Future*, 106–7.

33   Persson and Savulescu, *Unfit for the Future*, 123.

be welcomed and so it is surprising to read in another place they consider that there would be no serious crime in the world of moral technology, in part because criminals and potential criminals would be morally improved using whatever technology was available.[34] In spite of this idealism, it is extremely difficult to see in what ways people's altruism, concern for the poor, and reduced aversion to those of other racial and cultural groups can be so readily ameliorated using technological means of any description, let alone the means likely to be available in the foreseeable future. Additionally, a high level of moral awareness by the 'haves' will be necessary to avoid exploiting the 'have nots'.

Inherent within this endeavour is an assumption that a scientific approach to improving morality is able to determine what is desirable morally, or simply what is good as opposed to what is evil. It is one thing to argue that criminals will be prevented from continuing to act out their criminality, but who determines what constitutes the scope of criminality? One imagines it will be those with power in society. If these happen to be scientists, in what way will their science provide a guide to altruism, to appropriate behaviour on the battlefield or in business, or to resources to be devoted to the elderly? In the absence of such guidance, there will be no way of determining how technological prowess is to be utilized.

The concerns expressed by these writers are valid and greater note of them would assist greatly across every country in today's world. There are no easy answers and it is encouraging to see bioethicists attempting to come to grips with them. But is a technologically based approach the basis of hope or is it ephemeral? Is it a distraction or should neuroscientists direct their considerable efforts to find ways in which the moral wisdom of individuals and groups can be enhanced? After all, if our predicament is dire and if it can – theoretically – be assuaged technologically, concerted efforts should be directed at finding which chemicals, hormones and neurotransmitters will bring about the desired results.

---

34   J. Savulescu, T. Douglas, and I. Persson, 'Autonomy and the Ethics of Behavioural Modification', in Akira Akabayashi, ed. *The Future of Bioethics: International Dialogues* (Oxford: Oxford University Press, 2014), 91–112.

These possibilities stand at the border between bioethics and science. If any projects along this line sound unlikely, it may be because they are unlikely. Or it may be because we are not taking moral technology seriously. No matter what our doubts over this form of enhancement, we should not overlook our dependence upon technology in many instances where there is a discernible pathology (see chapter 2).

Finding a theological context for moral bioenhancement

In normal life we look favourably on enhancement. We routinely enhance someone's work or life prospects; it is far better to be provided with opportunities than be denied them. It is far better to have an adequate diet than an inadequate one, to have good living conditions than poor ones, to live a moral life as opposed to an immoral one. Christians as much as anyone else welcome enhancement in any of these senses. Why then may we be dubious about morally enhancing an individual or even a whole population technologically? What is it about technological intrusion that worries some of us?

We freely accept numerous intrusions into the human body: vaccines, surgery, and drugs to control blood pressure, elevate mood, regulate heartbeat and control movements. Evidently it is not these that worry us, even though some of them influence brain activity, and even though many of them are accompanied by unwanted side effects. We accept them because we believe they will assist us to live our own lives as the people we know ourselves to be. And so is moral neuroenhancement of a different character from any of these?

A core assumption is that we recognize what it is about our lives that constitutes normality. I think I am normal, but am I? This is sufficient for most people most of the time. Others also consider us to be normal. But what if our normality, what we consider to be the 'real me' is drug dependent? This is precisely what some people on Prozac have claimed when taken off Prozac: 'I'm not myself anymore'.[35] The perception of normality may therefore be changeable and fragile, dependent more on circumstances and context than we generally assume. It is within this framework that we have to consider where drugs, let alone education, counselling or pastoral care, fit in.

Some adopt a precautionary stance,[36] and express apprehension at what may be the unknown and possibly negative outcomes of biomedical technology. After all, the future is always unknown, and when a procedure is relatively untried the chances of something going awry is a significant factor to take into account. Therefore, so the argument goes, it is preferable to stick to the known and argue against the adoption of the technology in question. Unfortunately, this by itself does not provide a mechanism for critiquing the procedure, let alone implementing it or an alternative. Additionally, it does not attempt to grapple with the issues theologically, since precaution per se is not an outworking of Christian imperatives.

It is not unusual for Christian commentators to concentrate on the manner in which biomedical technology has the potential for changing humans in ways never previously contemplated.[37] This is the 'brave new world' that Aldous Huxley and many others have sketched, a world that appears to pose immense challenges for Christians since it gives the impression of wresting the control traditionally ascribed to God by placing it in the hands of fallible humans.[38] Moreover, these are the hands of scientists, and can scientists be trusted with such profound responsibilities as enhancing human traits, let alone ones aimed at enhancing moral sensibilities? Reminders of the foolhardiness of a science-as-saviour mentality are timely. Negative as this is, it constitutes a first step in laying the foundations for

---

35   P. D. Kramer, *Listening to Prozac* (New York: Viking, 1993).
36   Robert Song, 'To Be Willing to Kill What for All One Knows Is a Person Is to Be Willing to Kill a Person', in B. Waters and R. Cole-Turner, eds, *God and the Embryo: Religious Voices on Stem Cells and Cloning* (Washington, DC: Georgetown University Press, 2003), 98–107.
37   See Gerald P. McKenny, 'Technologies of Desire Theology, Ethics, and the Enhancement of Human Traits', *Theology Today* 59/1 (2002), 90–103.
38   Jones, *Designers of the Future*; Jones, *Manufacturing Humans*.

a more holistic and relational approach to human welfare. But it is only a first step. Remaining at this general level will never enable Christians to contribute usefully to ongoing debate.

The next step is to analyze neuroenhancements in terms of specific procedures and this is where scientific and ethical considerations enter the picture. Precise scientific control of the brain is an illusion, on top of which there is the ever-present problem of side effects. The problem of addiction is intimately bound up with the mechanisms of action of some of the dominant neuroenhancing drugs; one cannot have the enhancement without the addiction (see chapters 2 and 5). This has a direct bearing on the speculative vistas served up in this area, since they need to be tempered by empirical reality – what is and is not scientifically possible today and in the foreseeable future, as opposed to what might eventuate but for which there is no evidence of any description. It is also crucial to ascertain the balance between harms and benefits, and curing and caring, and whether the procedures demonstrate justice and concern for the poor, or fairness and neighbour love. These are primarily ethical questions, and while theological input has a contribution to make, it is just one ingredient within a broader set of queries. On the basis of the state of the science, the current consensus is one of concern, since measures taken to morally enhance individuals are accompanied by far more negatives than positives.

Beyond these immediate considerations, it needs to be asked whether biomedical technology is changing our conception of what constitutes the good life – or in Christian terms the life of faith? What is the moral life and what place is there for God's purposes and human responsibility? Jesus repeatedly stressed that what matters is what we are like as people rather than how we appear superficially (Matt. 23:1–12, 23–28, 33; Mark 7:6–8, 17–23). Our motivations and drives are central to who we are, since it is these that lead to the sort of choices we make and the directions we take. It is our motivations and drives that determine how we treat others and whether or not we put their interests before our own (Phil. 2:3–11; Eph. 4:29). How do we interact with the communities of which we are part? What are our priorities? Answers to these questions demonstrate the sort of people we are – empathic and loving or arrogant and self-centered, humble and sacrificial or boastful and resentful, hospitable and kind or unwelcoming and belligerent. The positive characteristics are formed through a lifetime of decision-making, all of which impacts our brains and neural organization (unfortunately, the same can also be said for the negatives). In this way the moral life (the spiritual life) is created and moulded. There are no shortcuts to moral behaviour.

This is not to argue that there is never any place for any technological intrusion into our brains. However, it is far easier to modify people in superficial ways than to transform what they are and what they stand for. Claims that healthy individuals can be morally changed for the better by changing one element of brain function is a form of reductionism that is misleading; it will never eventuate in this simple clean way.

In order to take these thoughts further consider the response given by Jesus to a lawyer, who wanted to know which was the greatest commandment: '"You shall love the Lord your God with all your heart, and with all your soul, and with all your mind." This is the greatest and first commandment. And a second is like it: "You shall love your neighbour as yourself." On these two commandments hang all the law and the prophets.' (Matt. 22:35–40)

Is there a place for moral technology in bringing about love like this? Is there any way in which we could envisage using technology to enable people to love God and those around them? Consider the following individuals.

> Jan is committed to loving God and her neighbour but suffers from bipolar disorder. She cannot escape either the frenzied states or the depressive ones, although treatment is proving helpful. There are times sometimes lasting for weeks on end, when her functioning is very restricted and during these periods she has little thought for her commitment as a Christian. However, on other occasions she is energetic and excitable and is highly productive, and it is during these times that she appears to relish her commitment and shows love towards all around her. However, she is deeply troubled by the black episodes and by what she perceives as her lack of concern for others at those times, as well as her lack of interest in anything spiritual. She is treated with mood stabilizers, including lithium and sodium valproate. She is very grateful for this and within a year her condition has improved markedly.

This is an illustration of a disabling, pathological condition that is often successfully treated using drugs. The pharmaceuticals enable Jan to function relatively normally, and in this way assist her to love others and hence improve the moral framework of her life. They have assisted her to live in the way she wishes to live.

> Greg is also committed to loving God and his neighbour, but has become addicted to viewing porn on his computer. This does not touch every aspect of his life, unlike the case of the previous individual. Neither is this usually viewed as a diseased state, but it is seriously questioning the extent to which he loves God in every facet of his life. It also throws doubt on whether he loves all those around him when in his thinking he perceives some as objects to satisfy his lust. He is deeply concerned about this and does not wish to continue to be subject to this addiction. Currently treatment involves counselling and the assistance of support groups. But what if it proves possible to utilize drugs that act on the brain's reward circuits, and counteract this form of addiction? What role might there be for them against the background of Jesus's teaching?

One has to ask what it was that led Greg initially into viewing porn, since if this had not occurred the addiction would not have kicked in. This is where the moral problems commenced. And so even if drugs to counteract the addiction become relevant and can be advocated, there is no hint that they will have any relevance prior to the start of the viewing. Once again, therefore, as with individual Jan, their role will be in treating a pathological process. Useful as this might be, the moral questions lie beyond their use. The drugs do not improve Greg's morality, they simply help him cope with the immorality to which he has become addicted. This is far removed from the moral technology advocated by some writers.

> This is taken further by individual Steve who has no interest in the precepts of loving God and loving one's neighbour. He lives for himself and his own welfare. His aim is to build his own empire of wealth and privilege. He gives no thought to social issues, whether poverty or climate change, or the plight of refugees or ethnic cleansing. These are never allowed to intrude into his world of riches and contentment.

How are we to approach this behaviour if we consider it suspect and highly questionable morally? Where might technology enter the picture? On the premise that drugs will be found to improve moral precepts, it can be hypothesized that one could transform this individual into someone who now loves God and those around him, from an atheist into a believer. In the unlikely event that such a change could be effected would the end result be any different from the changes that can be wrought using psychological conditioning or possibly torture?

The resulting individual, Steve transformed, may give the appearance of conforming to certain external expectations, but he would not be a more moral individual. The moral technology would have failed to improve the stock of moral behaviour. It may even resemble the results of classic psychosurgery of the 1940s and 1950s, when aggressive patients were transformed into placid conformists – without the aggression but without any interest in life or in the activities that had once been central to their existence.[39] The central queries are how moral is the use of such technology, and who would determine that love of God and love of neighbour are to be dominant characteristics of the lives of those in society (in fact those with power may wish to eliminate God thinking from people's vocabulary). The controversial nature of such a proposal is all too obvious.

It is also worth returning to Jesus who was well aware of the contrast between external appearances and inner motivations. Towards the latter part of the Sermon on the Mount, he explicitly pointed this out.

> Beware of practising your piety before others in order to be seen by them; for then you have no reward from your Father in heaven. So whenever you give alms, do not sound a trumpet before you, as the hypocrites do in the synagogues and in the streets, so that they may be praised by others [...] they have received their reward. But when you give alms, do not let your left hand know what your right hand is doing, so that your alms may be done in secret; and your Father who sees in secret will reward you. And whenever you pray, do not be like the hypocrites; for they love to stand and pray in the synagogues and at the street corners, so that they may be seen by others. (Matt. 6:1–5)

Outward conformity is superficial and may actually be misleading. If ways will ever emerge of improving the response of people in giving altruistically

---

39 J. L. Hoffman, 'Clinical Observations Concerning Schizophrenic Patients Treated by Prefrontal Leukotomy', *New England Journal of Medicine* 241 (1949), 233–6; Elliot S. Valenstein, *Great and Desperate Cures: The Rise and Decline of Psychosurgery and Other Radical Treatments* (New York, NY: Basic Books, 1986).

to help others, they will also have to ensure that there is no desire to demonstrate to others how generous they are being. This goes well beyond simply 'doing the right thing' but also knowing why you are acting in this way and wanting nothing in return. The drug regime may also have to curb the longing to demonstrate to others the degree of one's altruism.

A different current debate is the influence of neuroscience on criminal culpability, including that of adolescents. In discussing several US Supreme Court decisions, Steinberg concluded that 'the law is concerned with how we behave and not with how our brains function'.[40] While not discounting neuroscientific evidence, his contention is that neuroscience complements behavioural findings rather than making them more real. This position is a salutary one for advocates of moral bioenhancement.

The attempt to transform people mechanistically is a manifestation of a quasi-religious faith that scientific knowledge is the only legitimate form of knowledge. The message of moral bioenhancement is that everything about human life is confined to the physical, including moral behaviour. The realism of any religious approach is discounted, and yet the realism is not to be readily dismissed. As the apostle Paul encountered the difficulties and strife involved in radical transformation of priorities and attitudes, he knew there was no ready answer in religious observances. In writing to the early Christian church in Rome, it is not difficult to feel the intensity of the struggle.

> For we know that the law is spiritual; but I am of the flesh, sold into slavery under sin. I do not understand my own actions. For I do not do what I want, but I do the very thing I hate. Now if I do what I do not want, I agree that the law is good. But in fact it is no longer I that do it, but sin that dwells within me. For I know that nothing good dwells within me, that is, in my flesh. I can will what is right, but I cannot do it. For I do not do the good I want, but the evil I do not want is what I do. Now if I do what I do not want, it is no longer I that do it, but sin that dwells within me [...] Wretched man that I am! Who will rescue me from this body of death? Thanks be to God through Jesus Christ our Lord! (Rom. 7:14–25)

---

40   L. Steinberg, 'The Influence of Neuroscience on US Supreme Court Decisions About Adolescents' Criminal Culpability', *Nature Reviews: Neuroscience* 14/7 (2013), 513–18.

For Paul there was no way out of this tension in his own strength. For him the way ahead lay in the power and direction provided by the risen Christ. The contemporary question is whether taking appropriate drugs could have assisted him in his inner battle. While the theoretical nature of this possibility is clear, it raises the question of what would be the consequences of looking for a technological solution to such moral questions, in Paul's time and also today.

This is a confusion of domains that has much in common with retail therapy. Buying clothes or a new house or a more expensive car in order to fill the void in one's life is the answer of retail therapy. Replacing a part of one's brain or modifying brain circuits in order to overcome moral struggles and act more morally is the answer of moral bioenhancement. This is what one might term 'existential neural therapy'. Its undergirding is scientism, that leaves no room for any non-scientific input. Attempts to 'inject' morality into an individual are flawed since moral behaviour develops and matures with time, as struggles are overcome and tensions are resolved. The wise individual has thought long and hard about ways of resolving moral predicaments, about means of approaching moral quandaries, and has learned from mistakes and errors of judgement. It is a process that builds on experience and that takes note of wise counsel from across many fields of human endeavour. Instantaneous answers have no part to play in building up a moral repertoire, which for those working within a Christian framework will rely heavily upon the Christian Scriptures and the writings of Christian scholars through the ages. As emphasized in previous chapters due note will always be taken of scientific data, and will inform approaches to individuals such as Jan and Greg.

In writing to the Christians in Galatia, Paul put before them two ways of living: the one, uninformed by a spiritual dimension and the other, the converse.

> Live by the Spirit, I say, and do not gratify the desires of the flesh. For what the flesh desires is opposed to the Spirit, and what the Spirit desires is opposed to the flesh; for these are opposed to each other, to prevent you from doing what you want. But if you are led by the Spirit, you are not subject to the law. Now the works of the flesh are obvious: fornication, impurity, licentiousness, idolatry, sorcery, enmities,

strife, jealousy, anger, quarrels, dissensions, factions, envy, drunkenness, carousing, and things like these. I am warning you, as I warned you before: those who do such things will not inherit the kingdom of God.

By contrast, the fruit of the Spirit is love, joy, peace, patience, kindness, generosity, faithfulness, gentleness, and self-control. There is no law against such things. And those who belong to Christ Jesus have crucified the flesh with its passions and desires. If we live by the Spirit, let us also be guided by the Spirit. Let us not become conceited, competing against one another, envying one another. (Gal. 5:16–26)

The basis of what he is arguing here is provided by the gospel of Jesus Christ, but is also worked out in ways that his readers could understand. There is no suggestion that this is an easy path, but it is presented to them as the preferable path and one that is open to them. The moral instructions are clear, but they have to choose. Nothing is foisted upon them. They are treated as adults, with responsibilities to both themselves and others within their community. The contrast between this and the quasi-scientific, technological approach is marked, and is an important consideration when assessing the attractions of moral bioenhancement.

The answer is not to reject outright technological interventions into the brain, since some are helpful and assist an individual to live as they seek to live (see chapter 2). These are to be welcomed. By the same token there is no simple way of transforming an immoral individual into a moral individual by manipulating that person's brain. Certainly, treat whatever is clouding that person's thinking and responses using technological means, thereby enabling the person to be a whole person. One may wish to call this moral biotherapy, but it is far removed from moral bioenhancement with its theoretical capability of providing a person with a preset moral repertoire. This is an abrogation of the responsibility built into those made in the image of God and with God-like attributes.

CHAPTER 7

Ageing and immortal bodies

The impact of biomedical technology on our bodies takes many forms. Many of these we accept gratefully, such as the inroads of surgery and the medical treatments available for an increasing array of conditions. While all these will not be as successful as we might wish, their goal is the alleviation of sickness and hopefully a return to good health. These are benefits we appreciate, and in the main they raise few major ethical issues or theological concerns. Unfortunately, these benefits are far from equally distributed throughout the world or even across any one country, and inequality of this order has a plethora of ethical connotations. Similarly, there are also issues at the peripheries, where tensions arise over providing expensive treatments to those in need, or to one group of patients in preference to another group, or to younger patients rather than older ones. These quandaries constitute the bread and butter of serious ethical debate and are discussed routinely in the bioethical literature.[1] However, in this chapter they are no more than a footnote to two topics pertinent to the body and the challenges confronting it. The first is the manner in which we approach ageing and whether there is a place within this for some form of biological enhancement. The second is the strange world of plastination, where dead bodies are dissected, preserved in plastic and then put on public display to appear as if they were alive, with life-like postures and features. Each of these in its own way challenges attitudes to the human body, the challenge in each instance coming from the possibilities opened up by biomedical technologies available at present or likely to become so in the foreseeable future.

---

1   A. V. Campbell, *The Body in Bioethics* (London: Routlege, 2009); A. V. Campbell, G. Gillett, and D. G. Jones, *Medical Ethics* (Melbourne: Oxford University Press, 2005); Jones and Whitaker, *Speaking for the Dead.*

Ageing

Dylan Thomas's famous poem 'Do not go gentle into that good night', was written as he watched, and lamented, the death of his father.[2] It has come to symbolize the almost universal abhorrence of human death and especially the course of the death of loved ones. For Thomas, the revulsion was linked to old age and the process of ageing and all that so often goes with it.

> Do not go gentle into that good night,
> Old age should burn and rave at close of day;
> Rage, rage against the dying of the light.
>
> Though wise men at their end know dark is right,
> Because their words had forked no lightning they
> Do not go gentle into that good night.
> [...]
> And you, my father, there on the sad height,
> Curse, bless me now with your fierce tears, I pray.
> Do not go gentle into that good night.
> Rage, rage against the dying of the light.

Writing over sixty years ago, Dylan Thomas was appalled at what lay ahead. But have things changed? Do we have to rage as much today? Can we in some way ameliorate the worst effects of ageing and are enhancement technologies beginning to conquer death? Put simply, are we wiser and better equipped than the wise men known to Dylan Thomas?

For some the answer is an unequivocal 'yes', and the most extreme manifestation of this is provided by transhumanists who prescribe various enhancements to stave off ageing and death, that has even been described as a 'bad lifestyle option,'[3] necessitating the task of finding means for con-

2   Dylan Thomas, *Collected Poems 1934–1952* (London: Dent and Sons, 1952).
3   A. Heard, 'Technology Makes Us Optimistic; They Want to Live', The New York Times Magazine (28 September 1997) 84–9. <http://www.nytimes.com/1997/09/28/magazine/technology-makes-us-optimistic-they-want-to-live.html?page wanted=all&src=pm> accessed 11 April 2013.

quering and ultimately defeating both ageing and death. Far less extreme, but far more ubiquitous, is the booming market in 'anti-ageing' products. According to McConnel and Turner:

> Whatever differences separate technological utopians and pragmatic purchasers of expensive over-the-counter skin regimes, they share a common sentiment: they fear ageing and death and seek ways to remain youthful and vigorous. They are afraid of the ravages of time and are wary of growing old in a society that prizes and pays for youthfulness and perfect bodies.[4]

In his assessment of the role of medicine in recent years, Verhey argues that the powers of secularism and the powers of medicine have conspired to 'medicalize death'.[5] He writes, 'With the eclipse of Bacon's God, the Baconian project shaped the practice of medicine in the twentieth century and into the twenty-first. The "most noble" end of medicine was regarded as the preservation of life. In pursuit of that end, no patient would be regarded as "overmastered" by disease'.[6]

How then can we begin to unravel some of the issues here and where do enhancement technologies enter the picture? Indeed, is there any role for enhancement initiatives?

What is ageing?

While it is deceptively easy to talk about ageing, it is important to disentangle ageing from lifespan. While we may want the opportunity to age, because we assume it entails maturity and the goods that go with it, we do not want the physiological processes of ageing that cloud and possibly ruin all the hopes we hold out for that longer lifespan. In other words, most

---

4   C. McConnel and L. Turner, 'Medicine, Ageing and Human Longevity', *EMBO Reports* 6/S1 (2005), S59–62.
5   Allen Verhey, *The Christian Art of Dying: Learning from Jesus* (Grand Rapids, MI: Wm. B. Eerdmans Publishing, 2011).
6   Verhey, *The Christian Art of Dying*, 38.

people appear to want to live longer, as long as the deleterious effects of ageing do not accompany increases in lifespan.

But is the phenomenon of ageing normal or is it by definition abnormal? If one considers ageing of the brain it is generally easy to distinguish between what many of us would describe as normal ageing (forgetfulness and difficulty in remembering names) and explicit dementia which is a clearly defined pathology. There may be some overlap between the two, but the processes going on in the brain are in the end readily distinguishable. They are not simply the two ends of a spectrum; all very old people do not end up demented.

What then about ageing in general? Three models of ageing are often distinguished, reflecting as they do different degrees of technological optimism.[7] The first considers that the maximum lifespan attainable is very close to what has already been achieved (accepting the notion that 'three score years and ten', give or take ten years or so, is not that far off the mark). The second accepts that treatment options are available and will increase lifespan, but biomedical interventions will not make humans immortal (perhaps centenarians will become far more the norm, again give or take ten years or so). The third takes seriously the prospect of virtual immortality (according to which we should not be content with anything less than a few hundred years as an initial goal).

No matter how much these models appear to veer from one another, all assume the potential efficacy of biomedical technologies to increase life expectancy, an assumption that is not universally accepted by those who point to the potentially devastating consequences of a host of clearly recognizable phenomena. These include lethal viruses, nutritional imbalances and excesses, and political instability that could lead, and in some societies are already leading, to actual decreases in life expectancy.[8] While each of these three ways of looking at anti-ageing can be considered to

7   E. T. Juengst et al., 'Antiaging Research and the Need for Public Dialogue', *Science* 299/5611 (2003), 1323; E. T. Juengst et al., 'Biogerontology, "Anti-Aging Medicine," and the Challenges of Human Enhancement', *The Hastings Center Report* 33/4 (2003), 21–30.

8   McConnel and Turner, 'Medicine, Ageing and Human Longevity'.

incorporate elements of enhancement, the underlying thrust in each case is quite different. I shall argue that the enhancement paradigm is less helpful than sometimes assumed.

More helpful in my view is an analysis based upon the following four categories: technologies directed towards overcoming the appearance of ageing; overcoming the accompaniments of ageing; decelerating the processes of ageing; and overcoming the processes of ageing. These can be considered as a continuum from the everyday to the highly speculative and revolutionary.

Category 1: To overcome the appearance of ageing

This is the best-known category that is used to varying degrees by most people; it is the stuff of ordinary life in our societies. This is the enhancement emporium beloved of pharmacies, with their ever-expanding ranges of cosmetic and related products. Alongside these are Appearance Medicine practitioners and cosmetic surgeons, with antidotes for dealing with wrinkles, encouraging skin rejuvenation and correcting the effects of skin laxity, employing approaches using Botox, dermal fillers, laser treatments and a variety of surgical techniques. Not content with a long and healthy life, we also want to look young. This is hardly surprising in a culture that celebrates youth and youthfulness.

It is not my intention in this chapter to assess the validity of any of these approaches, but it is incumbent that we enquire what is the goal in utilizing anti-ageing therapies like these and whether there may be limits to their use. Since the latter half of the twentieth century, there has been increasing focus on the body as a vehicle for identity and self-expression, leading to far greater emphasis upon the role of appearance and consequently the desire for self-improvement. The result is that, in the eyes of many, beauty is now a crucial indicator of social worth.[9]

9 R. Honigman and D. J. Castle, 'Aging and Cosmetic Enhancement', *Clinical Interventions in Aging* 1/2 (2006), 115–19.

The superficiality and slick commercialism behind much of the cosmetic industry, including cosmetic surgery, can be unnerving. Allied to this is the apparent deceptiveness of granting older people the perception of being younger than they are, while leaving the crucial processes of ageing largely untouched. Over against this, there may well be a positive side to these procedures since decreased indications of ageing, especially facial ageing, may improve emotional and psychological well being.[10] Consequently, it is not impossible that cosmetic surgery fits within a healing paradigm by virtue of its therapeutic benefits to an individual's self-esteem and confidence. It has even been claimed that cosmetic surgeons 'operate on the body to heal the psyche'.[11] Nevertheless, too much should not be claimed for cosmetic interventions, since they are confined to alleviating the symptoms of ageing as distinct from combating the fundamental biological processes of ageing.[12]

In this they stand alongside two other categories employed to combat the appearance of ageing: prophylactic and compensatory. The former, prophylactic, incorporates activities such as exercise and diet aimed at staving off the onset of physical ageing; the latter, compensatory, includes the use of products like Viagra and replacement hormones as in HRT (hormone replacement therapy), designed to re-invigorate 'failing' functions and revert to more youthful standards.[13]

These technological approaches may or may not warrant an enhancement designation, even though on occasion they may enhance quality of life. Ultimately, however, any improvements are limited and, if carried out with the goal of reversing the fundamental changes of ageing, are doomed to failure. Moreover, if too much is expected of them the result may be

10   J. E. Zins and A. Moreira-Gonzalez, 'Cosmetic Procedures for the Aging Face', *Clinics in Geriatric Medicine* 22/3 (2006), 709–28.
11   S. L. Gilman, *Creating Beauty to Cure the Soul* (Durham: Duke University Press, 1998), 25.
12   J. A. Vincent, 'Ageing Contested: Anti-Ageing Science and the Cultural Construction of Old Age', *Sociology* 40/4 (2006), 681–98.
13   Vincent, 'Ageing Contested'.

dissonance, as when an individual's surprisingly youthful looking face is accompanied by an ageing and dysfunctional body.

Category 2: To overcome the accompaniments of ageing

The aim here is to overcome the diseases that tend to be the almost inevitable corollaries of increasing years: arthritis, osteoporosis and dementia, and also those diseases which are the proximate causes of death in old age – cancer, heart disease and lung disease.[14] As average lifespan has increased, so has the prevalence of these disease states. Increases in lifespan through public health measures and medical advances have not ensured continued quality of life in the later years. However, most people do not want merely to live longer, they also long for continued health and an acceptable quality of life in these extra years.

This could be achieved by tackling specific disease states piecemeal, or more effectively by intervening in underlying ageing processes that make us vulnerable to the chronic ailments of old age. This is what Juengst and coworkers call 'compressed morbidity'.[15] People will not live unusually long by current standards, but will be more likely to reach their full lifespan and live their final years as healthy, active individuals, until their 'final, swiftly fatal, decline'. This approach does not seek to extend maximum human lifespan, but to extend the illness-free years of life.

The challenges enshrined within these simple statements are enormous. Take just one example: Alzheimer's disease and its accompanying dementia. The tragedy of this is well known, and its link to ageing is equally firmly established. The numbers of people suffering from Alzheimer's disease are staggering, and increasing very rapidly. In a recent World Alzheimer Report, Alzheimer's Disease International estimated there were 36 million people living with dementia worldwide in 2010, a number estimated to increase

14   Vincent, 'Ageing Contested'.
15   Juengst et al., 'Antiaging Research and the Need for Public Dialogue'.

to 66 million by 2030 and 115 million by 2050.[16] Nearly two-thirds live in low- and middle-income countries, where the sharpest increases in numbers are set to occur. After age sixty-five, the likelihood of developing dementia roughly doubles every five years. Another way of expressing this is to say that, while the overall Alzheimer prevalence has been estimated to be 1.6 per cent, the rate increases to 19 per cent in the seventy-five to eighty-four age group, and to 42 per cent in the eighty-five and over age group.

The prevalence is increasing because the population is ageing and people are living to an older age. Not only this, it is increasing in low- to middle-income countries at a higher rate than in high-income ones.[17] This is partly explained by increasing life expectancy in these regions as a result of better healthcare and the elimination of infectious diseases. One estimate is that by 2050, one in eighty-five people worldwide will be living with Alzheimer's disease.[18] About 43 per cent of prevalent cases need a high level of care equivalent to that of a nursing home. If interventions could delay both disease onset and progression by a modest one year, there would be nearly 9.2 million fewer cases of the disease in 2050 with nearly all the decline attributable to decreases in individuals needing high level care.

There is no ready answer, since these constitute some of the most fundamental challenges presented by neuroscience. While means of delaying the onset of the dementia and diminishing its destructive force are on the horizon, and are to be welcomed, optimistic talk about overcoming the condition may amount to little more than glib and dubiously ethical assertions. A world without dementia is nothing more than an ideal; one might say a delusion.

16   N. L. Batsch, M. S. Mittelman, and Alzheimer's Disease International, *World Alzheimer Report 2012* (London: Alzheimer's Disease International, 2012) <http://www.alz.co.uk/research/world-report-2012> accessed 23 June 2013.

17   R. N. Kalaria et al., 'Alzheimer's Disease and Vascular Dementia in Developing Countries: Prevalence, Management, and Risk Factors', *Lancet Neurol* 7/9 (2008), 812–26.

18   R. Brookmeyer et al., 'Forecasting the Global Burden of Alzheimer's Disease', *Alzheimer's & Dementia* 3/3 (2007), 186–91.

In this instance, enhancement technologies are better viewed as therapeutic technologies, where the goal is rectifying what has gone seriously wrong. There is no question of raising individuals to standards of performance way beyond what is normally achievable. It is the far more modest task of enabling them to function as we would want all older people to function – recognizing others, enjoying the company of others, adhering to the values and revelling in the interests they have always had, and contributing to the world in an appropriate manner.

Category 3: To decelerate the processes of ageing – increase in lifespan

Life expectancy at birth in Europe has dramatically increased over the past 150–160 years, a development that has proved revolutionary for social and ethical expectations and indeed for the ethos of medicine. This increase has been from as low as twenty-five years in the inner suburbs of industrial towns to around seventy-five to eighty years today.[19] The decline in mortality in the nineteenth century stemmed from improvements in the external environment, particularly sanitation, with a consequent reduction in infectious diseases, and also vaccination. This radical change in life expectancy means that we today are substantially enhanced in biomedical terms when compared with our nineteenth-century forebears. The use of science and technology to overcome ill health and untimely or premature death has become embedded in our psyche, leading to ever-increasing medical interventions to secure health and longevity.

From this it follows that the most efficient way to extend human life expectancy with currently available technology is to devote resources to reducing infant mortality in the developing world. However, does this by itself tell us anything about increasing lifespan?[20] The answer is 'no'.

19 S. Szreter and G. Mooney, 'Urbanisation, Mortality, and the Standard of Living Debate; New Estimates of the Expectation of Life at Birth in Nineteenth-Century British Cities', *The Economic History Review* 50 (1998), 84–112.
20 Vincent, 'Ageing Contested'.

Consider the following. Life expectancy of older people has changed comparatively little over the past 150 years. For instance, the increase in life expectancy for a child born in the US between 1860 and 1960 was thirty-one years, but the comparable increase for a sixty-year-old was one year.[21] In other words, most of the gains have been in childhood and young adulthood, through the control of infectious diseases and adequate nutrition. Those who made it past these crucial periods, and who were not killed by an infectious disease in adult life or by accidents or in wars, had a good chance of enjoying a lifespan comparable to that of many people today. This should serve as a warning for those with unrealistic expectations of the life-extending powers of biomedical technology. Consequently, it is important to distinguish between life expectancy and lifespan.

If lifespan is to be increased, it is imperative to decelerate ageing. To accomplish this, an effective anti-ageing intervention, or series of interventions, would have to be found. Were this to eventuate, one estimate is that it might prove possible to increase the mean and maximal human life span by about 40 per cent, that is, a mean age at death of about 112 years for Caucasian American or Japanese women, with the occasional individual living until around 140 years.[22] However, this would necessitate lifestyle choices, such as abstaining from activities that lead to early mortality or accelerate ageing and decline, including smoking, and excessive fat and salt intake.

One of the well-authenticated approaches is calorie restriction. It has been known for more than seventy-five years that a calorie-restricted diet extends the lifespan of a large variety of small animals.[23] For instance, in

21   A. R. Omran, 'Epidemiologic Transition in the US: The Health Factor in Population Change', *Population Bulletin* 32/2 (1977), 1–42.

22   R. A. Miller, 'Extending Life: Scientific Prospects and Political Obstacles', *Milbank Quarterly* 80/1 (2002), 155–74.

23   C. M. McCay, M. F. Crowell, and L. A. Maynard, 'The Effect of Retarded Growth Upon the Length of Life Span and Upon the Ultimate Body Size', *Journal of Nutrition* 10/1 (1935), 63–79.

rodents, the lifespan can be extended by up to 50 per cent.[24] Preliminary results from long-term experiments with nonhuman primates suggest that this is also the case with larger animals, possibly including humans. Prolonged moderate calorie restriction extends the lifespan of rhesus monkeys by around five years or 16 per cent – thirty-one years compared to twenty-six years. It also inhibits the onset of cardiovascular disease, cancer, Type 2 diabetes, and endometriosis without deleterious effects on reproduction or brain morphology.[25] Similarly, calorie restriction in mice delays brain senescence and prevents neurodegeneration.[26]

But there is a catch. Although the degree of diet restriction varies between studies, it generally involves providing animals with 70 per cent of the calories an animal would normally consume on its own. If one were to reduce the caloric intake of humans by this amount, the reduction would be from 2,500 calories a day to 1,750.[27] For many people this would be seen as a serious burden. There also appears to be a genetic component. For instance, Ames dwarf mice carry what some refer to as a 'longevity' gene, $Prop1^{df}$, and when given a calorie-restricted diet they live about 70 per cent longer than normal laboratory mice.[28]

It is evident, therefore, that attempts to date to decelerate processes of ageing have been of limited efficacy. While the life expectancy in whole societies has increased significantly as a result of medical and public health developments, once these are in place, further increases in the lifespan of those living in these societies have been limited in scope. The processes of ageing have eluded sophisticated technological approaches thus far;

24   J. Koubova and L. Guarente, 'How Does Calorie Restriction Work?', *Genes & Development* 17/3 (2003), 313–21.

25   J. W. Kemnitz, 'Calorie Restriction and Aging in Nonhuman Primates', *ILAR Journal* 52/1 (2011), 66–77.

26   S. Fusco et al., 'A Role for Neuronal cAMP Responsive-Element Binding (CREB)-1 in Brain Responses to Calorie Restriction', *Proceedings of the National Academy of Sciences* 109/2 (2012), 621–6.

27   M. Lane, D. Ingram, and G. Roth, 'The Serious Search for an Antiaging Pill', *Scientific American* 287/2 (2004), 36–41.

28   A. Bartke et al., 'Longevity: Extending the Lifespan of Long-Lived Mice', *Nature* 414/6862 (2001), 412.

enhancement – if that is the appropriate term – may lie in nothing more profound than taking careful note of our diets. While this lacks the 'wow' factor demanded by most brought up in technologically dependent societies, its message is a salutary one.

Category 4: To overcome the processes of ageing – with the aim of achieving immortality

Caution of this order seems light years removed from the claims and aspirations of those bio-gerontologists who explicitly claim that it will be possible to reverse ageing or create immortality.[29] In this case the aim is not merely to decelerate the fundamental biological processes of ageing but to overcome them entirely so that there is no maximum human life span. In theory a person could live indefinitely, a fundamental tenet of transhumanism. However, this is far from a universal position among bio-gerontologists, many of whom are intensely sceptical about such vistas.

Research in this area is largely hypothetical, and focuses on attempts to slow physiological processes that produce ageing at a cellular level. Any success here will necessitate a range of breakthroughs, such as 'identifying genes for ageing, limiting metabolic and oxidation damage, removal of worn-out or damaged cells, [control of] cell replication and protection, and [possibly the] use of stem cells for replacement therapies and combating the reduced efficiency of the immune system with age'.[30]

Many of these ideas are associated with the name of Aubrey de Grey, and his 'strategies for engineered negligible senescence' (SENS).[31] His research focuses on the accumulating and eventually pathogenic molecular and cellular side effects of metabolic ageing, and on how regenerative

29   For example, S. Shostak, *Becoming Immortal: Combining Cloning and Stem-Cell Therapy* (Albany, NY: State University of New York Press, 2002).
30   Vincent, 'Ageing Contested'.
31   Aubrey D. N. J. de Grey, ed. *Strategies for Engineered Negligible Senescence: Why Genuine Control of Aging May Be Foreseeable.* Vol. 1019, Annals of the New York Academy of Sciences (New York Academy of Sciences, 2004).

medicine (see chapter 8) can thwart the ageing process. On top of this, the aim is to restore vitality and function to adults by reversing the processes of ageing, not merely arresting their further progress. This would ensure that indefinite lifespan would be available to people already in existence (including those suggesting this kind of research) and not just future children.

I do not wish to take these ideas any further since they take us well beyond the vistas of realistic science (see chapter 8). The fascinating aspect of them is that they give the appearance of being scientific, and are portrayed as functioning within the domain of science. But by taking science into such speculative realms, they are transgressing the boundaries of scientific endeavours. If they are legitimately scientific they will function as hypotheses that are open to being tested and if necessary refuted. While one can argue that reversing many of the ageing processes is a hypothesis capable of being tested scientifically, and even worthy of being tested, it follows that one has to be open to refutation of the hypotheses. In other words, one should be open to the possibility that immortality may not be achievable; however, some transhumanists are not inclined to accept this. One cannot assume technologically engineered immortality will ensue based on highly speculative scenarios.

Quite apart from this, how far could this take us? Potential immortality would have no effect on disease states or potential injury – a person may be ageless but could still be hit by a car or succumb to infection. In all probability ageless beings would have to cope with most of the mundane challenges with which mortal beings currently have to contend. And what about those living in the two-thirds world, for whom an increase in life expectancy from forty to fifty would be revolutionary and relatively easy to achieve? Perhaps it would be better to give ten more years of life to a forty-year-old in the developing world, than attempt to give 200 years or so to an eighty-year-old in the developed world. We dare not ignore questions of distributive justice.

Enhancement, anti-ageing and mortality

Where have enhancement technologies brought us? If one takes enhancement to encompass an extension of abilities beyond the normal or usual boundaries, the sort of boundaries that might be thought of as characteristic of the species, we can begin to appreciate how it might be applied to the enhancement technologies and ageing. Let me return to the four categories.

The first one attempts to overcome the appearance of ageing. Here the aim is to enhance the individual in a general sense, utilizing technological approaches. However, looking younger than your years, or reverting to what you hope was your appearance a few years previously, is enhancement in a very weak sense. It may be crossing a boundary, by going backwards in time in order to feel better or to increase one's self-confidence, but this is hardly a substantial boundary. Ageing is not being defeated in any major way. Ethically, this poses few problems, as long as there is not too much self-delusion and it does not waste undue financial resources.

The second category has as its goal overcoming the accompaniments of ageing. This is the realm of therapeutic technologies rather than enhancement technologies. However, a by-product of the therapy will be a distinct improvement in health, individuals who feel much better and also younger than their chronological age. In this sense, their lives have been enhanced. In this very loose sense enhancement is to be welcomed, although the challenges in some of these cases are daunting.

The third area has as its goal an increase in actual lifespan alongside an increase in life expectancy. Both are illustrations of enhancement, of whole communities in the case of increasing life expectancy, and of individuals in the case of life span. The former is truly transformative and, as we have seen, is brought about by eradicating infectious diseases, in some instances by utilizing what by today's standards is low-level technology. Increasing the life span of individuals is surprisingly problematic, with the one proven amenable contributory factor – restricting calorie intake – once again being very low-level technology. While this may not be the whole story, these are enhancement technologies in a weak sense. These technologies are to be welcomed.

The fourth category, overcoming the processes of ageing and ushering in immortality, introduces enhancement technologies in a strong sense. As has already emerged, this is by far the most speculative category and the one demanding the greatest amount of faith in the hypotheses and vision of its advocates, alongside the technological capabilities of scientific approaches required to bring this vision to reality. This is advanced speculation that has more to do with scientism than with science.

The fundamental query is whether the quest for physical immortality is misguided. Campbell introduces a hint of realism when he comments, 'We may seek a permanent habitation on earth, but the crumbling and illegible tombstones of bygone ages [...] bear ample evidence to the folly of such dreams'.[32] This is not to argue against efforts at increasing life expectancy. There is no virtue in dying at the age of seventy if one can live a productive, healthy and fulfilled life to the age of ninety or 100, and there is certainly no virtue in dying at the age of ten, twenty or thirty from eminently eradicable infectious diseases, malnutrition or genetic conditions.

A longing for some form of human immortality is a longing for more of the same: a longing for a prolonged earthly life, which itself generally has profound limitations. Meilaender has commented: '[...] a simple thirst for more (and more) life might seem to carry an unmistakable whiff of narcissism, for it is hard to imagine how we can act responsibly toward the generations that succeed us if we cling firmly (and desperately?) to our own continued youthfulness. Doing that would cause us to lose the shape that gives wholeness and integrity to our lives'.[33]

Mortality and ageing are inseparable. If we accept mortality, we are forced to accept one of its inevitable consequences, namely, ageing. If we reject the notion of ageing, we also reject the notion of mortality.

---

32 Campbell, *The Body in Bioethics*, 115–16.
33 Gilbert Meilaender, *Should We Live Forever? The Ethical Ambiguities of Aging* (Grand Rapids, MI: Eerdmans, 2013), 15–16.

Ageing and mortality – theological insights

The biblical writers had no illusions about old age. They recognized both its positive and negative features.[34] Old age is a blessing from God. In Proverbs, grey hair is seen as a crown of splendour that is attained by a righteous life (Prov. 16:31). In Ruth, God is viewed as renewing life and sustaining people in old age (Ruth 4:15). It may even be that the very long lives lived by the patriarchs were symbolic of the blessings of old age. Nevertheless, these positives have to be viewed against the backdrop of a range of negative ones. Barzillai in 2 Samuel 19:35 seems to be showing declining discernment, taste and hearing. Along similar lines the preacher in Ecclesiastes 12:3 describes losses of sight, teeth and physical strength with increasing age. The psalmist, writing in Psalm 90:10, considered that our span of seventy or eighty years amounts to little more than trouble and sorrow, although this may refer to far more than the latter years alone.

Besides these comments there is recognition of the wisdom of older people. In the Old Testament, wisdom and old age appear to have a special relationship. In this way, a corporate tradition was established and was brought to life for succeeding generations, who were to remember the days of old. This would be explained to them by older people, who had a special role to play in maintaining the nation's faith by recalling God's activity in the past. And so it was the patriarchs who were regarded as the epitome of wisdom in old age. It is also noteworthy that it was the ageing Simeon and Anna who greeted and proclaimed the Saviour (Luke 2:25–38). Respect for older people appeared to be one mark of a well-ordered society. In Israelite law, special provision was made for widows, while even adults were to honour their parents. Against this, disrespect for older people was a sign of chaos within society.

Coming through these and other instances is the centrality of hope no matter what changes occur throughout life. There is a continuity of identity regardless of what happens to an individual's body. This continuity

34   Jones, *Bioethics*, 214–15.

of identity emphasizes a broader continuity, from the past, through the present, and into the future. The 'real me' has been created by God and that relationship with God continues, no matter how frail or limited we may become. For Meilaender, hope is the virtue that sustains us and protects us from despair, since it leads us to desire something more than 'life's banquet'.[35] It also helps us come to terms with the realization that we will never attain a 'complete life' by ourselves, and it protects us from seeking a posthuman future of our own making.[36]

Another central theme is that of weakness. The temptation of modernity is to value most the powerful and capable, and the evidently successful. And yet the heart of the Christian gospel is the very opposite, characterized as it is by God's self-emptying and by the way in which he became vulnerable and weak in Jesus. Within a Christian framework, weakness becomes acceptable, becoming the norm and even the new way. Consequently, the weak, including the ageing, are to be highly valued within society.

Basic to these considerations is the limited life span of human beings. We are dust, and to dust we shall return (Gen. 3:19–20). The limited number of our days is regarded as a spur to the task of aspiring to wisdom (Ps. 90:12). It forces us to face up to the reality of our mortality, and to cast aside the illusion that we can go on living endlessly.[37] While some of the Old Testament writers explicitly accepted mortality, death was viewed in various guises. Some deaths were regarded as bad: a premature death (Isa. 38:10); a violent death (Amos 7:11); death without an heir (Eccles. 44:9). By contrast, death at a good old age and in peace (Gen. 15:15) and surrounded by one's children (Num. 23:10) was more acceptable, but still to be lamented.[38]

The evil of death in the Christian tradition is never underestimated or downplayed; it is never commended. The hope of the church lies in the resurrection of Jesus Christ with its 'celebration of God's victory over all

---

35 Meilaender, *Should We Live Forever?*, 36.
36 Meilaender, *Should We Live Forever?*, 36.
37 Jones, *Bioethics*.
38 Verhey, *The Christian Art of Dying*.

that hurts and harms [...] We will still grieve, but we need not grieve as if there were no hope'.[39] In light of these considerations Verhey concludes:

> A refusal ever to let die and the attempt to eliminate suffering altogether are not signs of faithfulness but of idolatry. And if life and its flourishing are not the ultimate goods, neither are death and suffering the ultimate evils. They need not be feared finally, for death and suffering are not as strong as the promise of God. You need not use all your resources against them. You need only to act with integrity in the face of them.[40]

A Christian conception of ageing and death takes account of the limitations of the human condition and the importance of being willing to accept these, not in a fatalistic way but in dependence upon the God who brought us into being and who sustains us and cares for us. This calls for patience and an understanding of the goals of human life and of our own lives as individuals. We are not responsible for our own existences; in this sense our lives have been given to us. No matter what the circumstances by which we came into being, we are the beneficiaries of this gift of life.[41] Consequently, each step along the way, including ageing and ultimately death, is part of this gift. It is incumbent upon us to do what we can to ameliorate ills and deficiencies, and in doing so to extend the length of life, but this ethos differs fundamentally from one that seeks by all means to overcome the givenness of life and its inherent limits.[42] Such a project also gives the impression that the present generation – those currently alive – should be the main focus of attention.

While both the transhumanist and Christian conceptions recognize that death is an evil, they part company over their remedies. For the one, with its faith in the unending ability of science, death is to be (has to be) defeated technologically. For the other the inevitability of death is accepted but transformed by faith in what God has achieved through the death and resurrection of Christ. For the one, efforts are to be directed at finding ways

39   Verhey, *The Christian Art of Dying*, 197–201.
40   Verhey, *The Christian Art of Dying*, 215.
41   Sandel, *The Case against Perfection*.
42   Meilaender, *Should We Live Forever?*

of prolonging life and defeating physical death, because this is the ultimate value. For the other, its inbuilt virtue of hope points to the expectation that we care for others rather than caring only for our own survival.

Plastination

In this section I am moving from ageing of the living to ways of preserving and displaying the dead. At first glance there may appear to be no relation between the two. However, in both cases the intrusion of technology has converted relatively simple processes into complex ones. Not only this; in their different ways both have aspirations to overcome the finality of death, the aim in both cases is to transcend what we are as mortal beings.

Plastination is a method of preserving human or animal tissues by replacing the tissue fluids with plastic. The specimens preserved in this manner are dry, odourless and durable, thereby making them easy to handle; the process retains the natural structure of the tissues. Using this process body parts and organs are routinely preserved in medical schools for the teaching of human anatomy. They are also used for research purposes. These uses raise no particular ethical or theological issues. However, what has emerged over the past twenty or so years is the preservation of whole bodies, the so-called 'plastinates'. If these were used for teaching and research within health science contexts, there may be no profound issues, on the assumption that they had been obtained in a legal and ethical manner (after fully informed consent). But that is not the case. Most plastinates appear in large public exhibitions, and are displayed as though they are alive and are participating in some activity – generally a sporting activity, playing chess, riding a horse or even having sexual intercourse. Besides these, others replicate a variety of artistic masterpieces from the Renaissance era. Beyond this they have been dissected to demonstrate muscles and nerves, organ systems and various planes through body regions. The effect is frequently dramatic and awe inspiring, and elicits reactions of wonderment at the

beauty and complexity of the human body (for some idea of what plastinates have been produced Google 'plastination images').

What are plastinates?

Since the plastinate is 30 per cent human tissue and 70 per cent plastic, it
raises the question of its relation to the 'normal' human body. Its largely
plastic composition means that it is an ambivalent entity. And yet it retains
many human features and much of the cadaver's physical substance; it also
retains marks of the individuality of the once living person.

This process represents a radical transformation of the human body,
since the preservation techniques are being used to give the impression
that individuals continue to live on in a dissected state. In plastination
exhibitions, of which *Body Worlds* was the forerunner and remains the best
known, these dissected plastinates are presented with a welcoming, almost
life-like visage, and serene facial expressions. The result is transformation
of the corpse into what has been described as the 'post-mortal body' to
give the impression that these dissected corpses are still alive.[43] Since these
plastinated bodies are two-thirds plastic, they have overtones of physical
permanence; in one sense they can be described as immortal. This so-called
'immortality' is technologically derived and constitutes a new category of
human body, neither fresh corpse nor decaying remains.[44] The artificiality of these bodies could be described as cyborgian, even if the machine is
non-functional.[45] Is it too fanciful to compare these elaborate plastinates
with cyborgs, part-man part-machine (see chapter 8)? The major distinction is that plastinates are dead, whereas cyborgs are, or will be, alive. The
artificial machine-like element in plastinates is dead; it is going nowhere
and will accomplish nothing other than to fulfil artistic and possibly edu-

---

43   PRNewswire, 'Anatomist Dr. Gunther Von Hagens Reiterates His Mission of Public
     Health Education to Press Corps in Guben, Germany', (30 November 2006) <www.
     prnewswire.co.uk/cgi/news/release?id=185453> accessed 20 September 2010.
44   Jones and Whitaker, *Speaking for the Dead*, 104.
45   Jones and Whitaker, 'Transforming the Human Body'.

cational goals. At the most they are non-functioning cyborgs (mixture of man and machine), and so while theoretically cyborgs may one day emerge to threaten living humans, plastinates never will.

The plastinate is more than a generic plastic model or even a model of an individual. It would not exist without the original human being, and yet the form of the plastinate – what it shows, the posture in which it is placed, the activity in which it is 'participating' – is the work of a mastermind, the master technician who has made these decisions and carried out this work. The resulting plastinate may be depicted indulging in activities never carried out during life, indeed activities the donors may not wish to have undertaken. This suggests that the appearance and apparent 'interests' of plastinates after death are a fiction, and it is this fiction that constitutes an additional barrier to viewing them as human. The aim is to make them into something new – to get away from the people they were during life, and create a new entity – based on a human template but increasingly artificial. As they take on a new visage, the memories of who they once were have been largely obliterated. There are discernible limits to the extent to which the artificial can intrude into human existence before core features of human life are lost.

Consequently, the end result is a conundrum, since while the plastinates are allegedly about the dynamic and living body, the newly constructed body is far removed from that of the original living individual. They represent their own category of living deadness, 'living lives' that fulfill the whims of their creator craftsman. This post-mortal world may be what the donors wanted for their mortal remains, and yet it represents far more the artistic world of their creator, an empire of someone else's desires.

The intention of *Body Worlds* is not to present dead bodies in their 'deadness'.[46] For Gunther von Hagens, the founder and mastermind behind *Body Worlds*, plastination has transformed bodies and in so doing has moved them beyond death. Von Hagens describes the plastinates as 'frozen

---

46 Kenji Skulstad, 'Body Worlds Draws Large Crowds – and Controversy', *Canadianchristianity.com* (4 July 2006) <www.canadianchristianity.com/cgi-bin/ bc.cgi?bc/bccn/1106/18body> accessed 12 October 2007.

in time between death and decay',[47] for they will never decompose, and thus have achieved a form of immortality. This is immortality, not in any spiritual sense, but of physical permanence; a post-Christian, secular form of immortality.[48]

One has to ask whether running throughout all of von Hagens' endeavours there is not an attempt to escape from the reality of death, by giving the impression that these cadavers are continuing to exist in much the same way as when they were alive. The artificial has triumphed, having eradicated decay and disease; it may even have attained eternal bliss. We can 'live on' as plastinates, with our post-mortal bodies. While this is not the ageless existence that transfixes transhumanists, plastinates appear to have attained their own form of everlasting existence, although this appears to be akin to the ongoing existence of some inanimate object.

Ambiguous nature of plastinates

These exhibitions have enticed a diverse array of scholars to engage with the human body. After all, dissecting human bodies is an exercise normally reserved for pathologists carrying out autopsies with the object of determining the cause of death. It exists on the edges of respectable society. This has always been the case in all societies, and it continues to be the case even though in many societies it has been accepted practice, for as much as 200 years in some cases. Over recent years the explosion of interest by the general public in viewing dissected bodies has opened up the human body to broad academic interest, and commentators on *Body Worlds* have come from a multiplicity of backgrounds – sociology, law, gender studies, art history, performance theatre, tourism, religious studies and communication

---

47   Linda Schulte-Sasse, 'Advise and Consent: On the Americanization of Body Worlds', *BioSocieties* 1 (2006), 369–84.
48   Megan Stern, 'Shiny Happy People: "Body Worlds" and the Commodification of Health', *Radical Philosophy* 118 (2003), 2–6.

studies. These and other contributors from the humanities have produced multiple, and sometimes conflicting, interpretations.[49]

What is interesting about these contributions is that most concentrate on what they regard as the transgressive nature of plastinates. Many of the dominant themes that emerge from the literature focus on the ambiguous ontological status of plastinates. They transgress the familiar categories with which we usually make sense of the world. These are: interior or exterior, real or fake, dead or alive, bodies or persons, self or other. By refusing to occupy this familiar binary terrain plastinates create tensions that our traditional concepts find difficult to encapsulate. This can be illustrated by reference to one of these areas.

It is the realism inherent in the use of human bodies rather than models that proves one of *Body Worlds'* greatest attractions, and one of its greatest marketing angles. This imperative to present and experience something 'real' is a peculiarly late modern desire, one in which reality is equated with immediacy, eschewing representation. Von Hagens strenuously distinguishes his exhibits from any kind of anatomical model, which, he claims has 'always been intellectually "regurgitated", and as such is always an interpretation [of reality]'.[50]

Because of this *Body Worlds* is heavily marketed as the exhibition of 'real' human bodies. The attraction for most exhibition-goers lies in the reality, that in viewing real bodies in this way we/they are approaching the forbidden. As long as visitors believe the plastinates are real bodies they experience the thrill of taboo breaking and voyeurism. This is fortunate for *Body Worlds*, for while the realness of the plastinates may be the lure, the plastinates themselves hardly look like real bodies, in part due to the

---

49 Mike R. King, Maja I. Whitaker, and D. Gareth Jones, 'I See Dead People: Insights from the Humanities into the Nature of Plastinated Cadavers', *Journal of Medical Humanities* (2013), Advance online publication, 22 March 2013, doi: 10.1007/s10912-013-9230-z; J. D. Lantos, ed. *Controversial Bodies: Thoughts on the Public Display of Plastinated Corpses* (Baltimore, MD: Johns Hopkins University Press, 2011).

50 Gunther von Hagens, 'Anatomy and Plastination', in Gunther von Hagens and Angelina Whalley, eds, *Anatomy Art – Fascination beneath the Surface* (Heidelberg: Institute for Plastination, 2000), 11–38, 33.

plastination technique, and in part due to the artistic licence of the plastinators. Moreover, as previously outlined, the plastination process itself drastically reduces the amount of original human tissue in the plastinated body to a fraction of the whole.

It has been claimed by one writer that the bodies are chemically, surgically and artistically modified to such a degree that the intrusion of the artificial makes them 'hyper-real'.[51] Hence, it is possible to argue that what *Body Worlds* presents are not real bodies but representations of real bodies. The donated body is manipulated and transformed in such a way that the end result is an artificial representation of perfected nature. However, this is not the impression the exhibitions wish to convey.

Plastinates bring us face to face with the nature and status of the contemporary body, that is both dead and alive – neither natural nor artificial, but some fusion of the two. Here the categories of natural and artificial map onto real and fake, and the distinctions within each set have become arbitrary and obsolete. In modifying bodies in this way the aim appears not to be imitation of the pure and natural with artificial modifications, but to modify natural bodies into the form of artificial representations of the natural. The bodies are no longer real or fake. The 'authentic' body, it is claimed, has become the plastinated body rather than the unmodified natural decomposing body.[52] However, in doing this, the concept of authenticity has been changed.

Visitors experience plastinates as both real and unreal, so that they are unsure what it is they see – a corpse, a statue or something in between.[53]

---

51   Jane Desmond, 'Postmortem Exhibitions: Taxidermied Animals and Plastinated Corpses in the Theaters of the Dead', *Configurations* 16/3 (2010), 347–78; Megan Stern, 'Dystopian Anxieties Versus Utopian Ideals: Medicine from *Frankenstein* to the *Visible Human Project* and *Body Worlds*', *Science as Culture* 15/1 (2006), 61–84.

52   T. Christine Jespersen and Alicita Rodríguez, 'Forced Impregnation and Masculinist Utopia', in T. Christine Jespersen, Alicita Rodríguez, and Joseph Starr, eds, *The Anatomy of Body Worlds: Critical Essays on the Plastinated Cadavers of Gunther Von Hagens* (Jefferson, NC: McFarland & Co, 2009), 166–75, 170; von Hagens, 'Anatomy and Plastination', 20.

53   Tony Walter, 'Plastination for Display: A New Way to Dispose of the Dead', *Journal of the Royal Anthropological Institute* 10 (2004), 603–27.

Even though they may go to *Body Worlds* looking for authentic bodies, this is not what they find. Von Hagens enthusiastically advocates the notion of post-mortal existence based in ongoing physicality. In doing this he has moved beyond a mere technique, transmuting it into a means for satisfying the aspirations of the donors for a form of 'immortality'. This move, however, has taken him, and the *Body Worlds* exhibitions, into profoundly ambiguous territory.

If this is immortality it is a secular, material form of immortality. However, the impression of imparting biological life after death is illusory and artificial. While the plastinates are posed to suggest movement or activity, they are in fact eternally static, merely mimicking vitality. This 'post-biological existence' is no more than a synthetic afterlife; they may appear immortal, but they lack almost all identification with the person who once lived. The trace that remains of those who were once alive is not of values, attitudes or ideas. It is a static bodily remnant that itself is likely to bear little resemblance to the external bodily features by which the individual was known and by which he or she would usually be remembered. Consequently, the plastinates occupy a problematic intermediary position between our commonly held conceptions of dead and alive. At its best *Body Worlds* provides a reflective space to meditate on mortality, since it has made the dead body the most accessible it has ever been for those outside the anatomical and health sciences. Yet one can question what it is that has been made accessible.

Plastinates do not fit neatly, much less exclusively, into any recognizable categories of humanity, as demonstrated by all commentators who grapple with a range of dichotomies, ambiguities and contradictions. These are summed up in the realization that plastinates are metaphysical misfits, since they disturb and challenge our usual analytical categories. They are variably seen as controversial, intriguing, disgusting, fascinating, disconcerting, confusing or awe-inspiring to those who view them.

But have they escaped the limits of mortality? Are they everlasting? Or in acquiring some form of everlasting existence have these plastinates ceased to be human? One observer expressed his response with these words:

Each of [… the] corpses is, at one level, a perfect human specimen that is a real privilege to observe at close quarters. And yet, the absence of a personality, friends, family
and history leaves a gaping and eerie vacuum that forcefully calls into question what
it is to be human and reminds us of what few of us like to dwell on – our mortality.
They are bodies with no soul.[54]

Everlasting existence may, in one sense, have been achieved, but only by
sacrificing their human core. The replacement of dead human tissue with
plastic leaves nothing more than plasticized remnants of what was once
human, fascinating and yet perplexing, uplifting and yet troublesome.
We seem to have created a new form of existence, perhaps plastinates is
indeed an appropriate term, and yet we have not come to grips with what
this means.

In April 2009 I went to the *Body Worlds* exhibition the *Mirror of Time*,
in London, with its emphasis upon human development and ageing. As a
Christian and an anatomist I wanted to see how I responded to the exhibition from these two perspectives. Of course I cannot isolate the one from
the other, since they inform each other. However, it is the perspective of
the Christian anatomist that is relevant in the present context.

As I looked intently at some of the plastinates, I was fascinated,
intrigued and perplexed. As I came very close to them I was forced to
ask just what it was that I was looking at. Certainly, I was in awe of the
anatomy and of the exceedingly high standard of the dissection. The anatomical detail was masterly, on two counts. The first was the skill of those
responsible for the dissection and subsequent plastination, and the second
was the reminder of the exquisite arrangement of the nerves, vessels and
organs that make the functioning of the human body a reality. However,
there was even more to the experience than this. I had to remind myself
time and again that I was coming face to face with dead individuals. I was
in contact with death. And yet the serene expressions on many of their
faces were benign; they seemed to be enjoying the experience of being dead.
They gave the impression of being content with their present (deceased)
state. They had reached a blissful end, perhaps they were in a 'post-mortal

---

54   E. H. Nicholls, 'Selling Anatomy: The Role of the Soul', *Endeavour* 26/2 (2002), 47.

heaven'. It would not have taken much to convince me they were so much better off than those who are decomposing, or were reduced to ashes a long time ago in a crematorium. And yet ...

This is where the ambiguity of this form of transcendence has brought me. We are more than just our bodies, to both us and others, since we cannot be reduced to our bodies alone. This is almost self-evident in life. But neither can this be the case after death. In this sense plastinates, masterly dissections that they are and masterly anatomical 'replicas' that they are, are also deceptive if we think they are more than cadavers. Being preserved after death as a plastinate has no advantages to decomposing as a corpse (unless one wants to learn about the anatomy of the human body and/or appreciate the aesthetic beauty of the body's organization).

Elements of a Christian response

The preceding discussion has demonstrated that the manner in which plastination has been employed in many of the public exhibitions raises profound questions of immediate theological interest. Many of the initial responses by Church representatives have been cautionary or negative. Church spokespeople tend to object to shows of this nature, and sometimes, even call for them to be banned.[55] The reasons for this reaction are likely to be on the ground that the body is sacred, and that these shows are dishonouring the body, or that the dignity of human beings is being compromised. On occasion the claim is made that a moral and ethical line has been crossed, since displays of bodies amount to a desecration of all we hold dear.

Plastination is a very modern phenomenon, even though there are historic precedents in a variety of anatomical exhibitions since the eighteenth century.[56] There is no way in which any biblical writers could have

---

55 Linus Dolce, 'Injustice Perpetrated on the Dead', *The National Catholic Bioethics Quarterly* 10/4 (2010), 667–76.
56 Jones and Whitaker, *Speaking for the Dead*, 90–1.

possibly envisaged anything of this nature. But there may be clues in the views of the biblical writers on the dead human body.[57]

A first element stems from examples in both Old and New Testaments of the high view held of the dead body. An example of this is found in Amos, who specifically separated out for condemnation the crimes of one group of people who, not content with marauding, pillaging and killing, unleashed their venom on the body of one of their enemies. Having killed the king of Edom (Amos 2:1–3), they burnt his bones to ash. Not content with killing him, they desecrated his dead body thereby undermining his integrity as an individual.[58]

Another instance of the significance ascribed to the dead body is provided by Joseph who, prior to his death, had his relatives promise to take his bones with them to the land of Canaan when they were finally able to leave Egypt (Gen. 50:22–26; Exod. 13:19). Dead though he would be, Joseph did not want his mortal remains to be left in Egypt, the land of captivity. This may have been symbolic, and yet it strengthens the notion that the dead body is sufficiently important to require commitment on the part of others. In the New Testament, we find that, following Jesus's death, his followers carefully and sacrificially tended his body (Matt. 27:57–61; Mark 15:42–16:2; Luke 23:50–24:1; John 19:38–42). They considered it inappropriate to leave his body on the cross, especially as this would have meant leaving it there over the Sabbath Day. Joseph of Arimathea ensured that Jesus's body was laid in his own new tomb, while a number of his followers, including Nicodemus and Mary Magdalene, were concerned that the body was anointed with spices and bound according to Jewish custom. While there are many cultural factors here, there is no hint that Jesus disapproved of their actions. There was nothing untoward in looking after his dead body in this way. His followers may have underestimated the likelihood of his resurrection, but that was another matter. What is

57   Jones, 'The Human Body: An Anatomist's Journey from Death to Life'.
58   D. Gareth Jones, 'The Human Cadaver: An Assessment of the Value We Place on the Dead Body', *Perspectives on Science and Christian Faith* 47 (1995), 43–51.

encountered in these instances is clear recognition that the dead body is to be treated with respect.

This conclusion is not surprising since there is no suggestion in the Bible that human beings can exist apart from the body, even in the future life after death. In unequivocal terms, Paul enunciated the point that the resurrection is a physical one (1 Cor. 15:42–52; 1 Thess. 4:13–18), a belief foreshadowed in the Old Testament (Dan. 12:2). A biblical view, therefore, militates against any idea of humans existing apart from some bodily manifestation or form of expression.[59] The mortal body we now know will be transformed into a resurrection body, a form of spiritual body, which while not identical to our present material body has sufficient similarities to it to warrant the term 'body'.[60]

Jesus's own resurrection body serves as the only guide we have to this (Luke 24:12, 31), with its recognizably human and personal features but also its ability to pass through material objects. Bearing this similarity in mind we can go further and argue that respect for the dead body now foreshadows respect for the resurrection body in the future. While in no way suggesting there is a close parallel between the two, there would appear to be connections. A willingness to desecrate or devalue the dead body shows a disregard for what that person may become, as much as it shows a disregard for what that person has been. While it is not our prerogative to judge what any person may be like in a future dimension, it is our responsibility to provide support and protection as far as we know how. Any Christian perspective, therefore, is to take account of this future dimension in determining how a dead body is treated in the present, taking account of the notion that this present life is preparation for a future one. An element within this perspective is that the prior wishes of the deceased are

---

59 B. O. Banwell, *The Illustrated Bible Dictionary* (Leicester: Inter-Varsity Press, 1980), 202–3.

60 Green, *Body, Soul, and Human Life*, 140f; Peter Lampe, 'Paul's Concept of a Spiritual Body', in Ted Peters, Robert John Russell, and Michael Welker, eds, *Resurrection: Theological and Scientific Assessments* (Grand Rapids, MI: William B. Eerdmans, 2002), 103–14; Wright, *The Resurrection of the Son of God: Christian Origins and the Question of God*.

respected as far as possible, since our bodies constitute the one common strand between what we are now and what we may become.

Hence, while a Christian response may have similarities to many general ethical stances, it goes further by recognizing that the dead body serves as a link between that person in the past and future. It imparts a future orientation to add to the past orientation of general ethical perspectives. The body itself is an inadequate token of these dimensions, but it is all that remains. It is a reminder of the greater ongoing dimensions of human existence, and also of the reality of our limitations and needs, as well as of our mortality. We shall all die and be like this one before us who is now dead. Respect for the dead body reminds us, not only of the significance of the one who has died, but also of the significance of all human life. Consequently, to value the dead body is to value the person, and to see that person as one who mirrors God. To devalue the dead body is to devalue those still alive, and also to call into question the purposes and intentions of God in creating people in his image.

But what amounts to devaluation of the dead person? Does plastination do this? What about far better known examples like dissection, along with organ transplantation and donation? Perhaps those church spokespeople were right all along: we should oppose any public display of bodies. Or are they?

There can be no doubt that dissection amounts to mutilation of the body. But mutilation does not inevitably amount to desecration. One crucial factor prevents this, and this is the role played by informed consent.[61] It is this that transforms an act of desecration into an act of respect, simply because what is then done with and to the body is a fulfilment of the considered wishes of the person when still alive. Undergirding these wishes is the person's altruism, the gifting of their dead body to others. The aim is to ensure that their body is put to good use, namely, education and/or research,

---

61   Department of Health, *The Removal, Retention and Use of Human Organs and Tissue from Postmortem Examination* (London: Her Majesty's Stationary Office, 2001); Retained Organs Commission, *A Consultation Document on Unclaimed and Unidentifiable Organs and Tissue, a Possible Regulatory Framework* (London: National Health Service, 2002).

so that people in future will benefit from this gift. Their life has come to an end, but their mortal remains may be able to assist others in some way.

This is a crucial perspective both ethically and theologically, since it enables the dead body to be put to uses that would not otherwise be possible. The only way in which this perspective of respect and a dissecting ethos can be held together is via altruism.[62] In these terms, the sole justification for dissection within a Christian perspective stems from the altruism of the living, in that the person when alive decided to gift her body to a medical school in order to be used in a certain way following her death. The specific Christian thrust within this principle is that the supreme model available is that of Jesus himself, in giving up his own life for others. To give one's life for one's friends is ethically commendable, and to do it for those who are undeserving is the height of altruism (John 15:13). In first becoming a human being, and then in giving up that life voluntarily for others, Jesus demonstrated in unequivocal terms the characteristics of a life of humility rather than of arrogance or conceit (Phil. 2:3–8). The gift of one's body after death in no way matches altruism of this calibre; it is, therefore, a limited form of altruism. Nevertheless, it is imbued with the essence of altruism.

But where does the public display of plastinated bodies and body parts fit in? Informed consent is as crucial as ever, but is altruism as strong? Is the body being given to others for good purposes? What benefits emerge from the public displays of these so-called plastinates? The three that emerge in these discussions are education, art and entertainment. Of these it is the third that poses the greatest problems: it can be considered degrading since it demeans all that those bodies have stood for as integral facets of human persons capable of relating to each other and to God. In practice, of course, there is undoubtedly a mixture of all three possible benefits, with education, art and entertainment jostling for pride of place. The ratio of each to the others is what has to be assessed before one can arrive at a reasoned ethical and theological judgement. My conclusion is that greatest

---

62   W. F. May, 'Religious Justification for Donating Body Parts', *Hastings Center Report* 15/1 (1985), 38–42.

emphasis should be placed upon education, lesser emphasis upon art, with the entertainment side being of limited consequence.

But is there something inherently different about plastination? My conclusion is that it objectifies bodies in a way not encountered in any other post-death procedure. The plastinator is part-manufacturer, making changes to the plastinate in a way never found in other procedures like dissection or organ transplantation. The rationale for this is entertainment, and therein lies the potential for lowering human dignity and for demeaning these human remains as beings made in the image of God.

This is the crux of a theological response to plastinates, since the 'immortal' specimens have become what they are through the creativity of human beings. But in order to achieve this status, they have been torn asunder from their human roots. They are not human beings with the gift of longevity let alone agelessness. They have been created in the image of a human being – their plastinator creator – with the goal of entertaining the masses through a commercial enterprise. As a general principle, being buried or cremated is more in keeping with a Christian ethos.

CHAPTER 8

Increasing dependence upon technology

In the preceding chapters I have scoped a diverse array of issues, from the artificial reproductive technologies to neuroscience, from genetics to the technological enhancement of morality, and from ways of transforming ageing to the strange procedure of plastination of the human body. In each case, issues have arisen on account of the character of the technological intrusion into these processes. Some of them have been present reality; others have postulated ways in which the technology may be applied in the future. While there are notable differences between these groups, the dividing line between reality and speculation may be of lesser rather than greater importance. This is because all exemplify the extent to which increasing numbers of people are looking to technology to solve problems and even provide meaning for human existence by expanding the dimensions of human experience.

In this chapter I continue with this theme, and again aim to encompass the not so spectacular along with what some consider as the outlandish. I shall move from regenerative medicine, to cyborgs and on to post-persons, a gradation that for me signifies the transition from the realistic and therapeutic, to the speculative, although even here there may be therapeutic overtones in some instances. All in their own ways elicit considerable debate within bioethical circles and should not be ignored by theologians. This is not to suggest that they are of equal status or comparable importance, and yet they fit onto a continuum from the ordinary to the extraordinary. How do we distinguish between them beyond our own instinctive attraction to one or more of them or equally our disgust and revulsion at some of the possible directions?

All are topics that are calling out for Christian engagement. With this in mind I shall turn in the latter part of the chapter to consider what may be some theological challenges and the contribution theological input can make to our thinking when confronted by such novel possibilities.

Regenerative medicine

The term 'regenerative medicine' first appeared in the literature in 1992 when it was depicted as a hypothetical future technology capable of revolutionizing clinical treatment.[1] The hypothetical took on a far more realistic façade in 1998 with the isolation of embryonic stem cells.[2] At much the same time the possible clinical significance of the growth potential and pluripotency of these stem cells was appreciated. As pointed out by King et al.,[3] this opens the way for damaged tissues and organs to be regenerated by the insertion of stem cells, stimulation of endogenous stem cells, or transplantation of tissues or organs grown *in vitro* from the patient's own stem cells.[4] The therapeutic dynamic behind these techniques is that they will radically advance the treatment of diseases from diabetes to spinal cord injury, and from Parkinson's disease to Alzheimer's.

A description like this should raise no ethical or theological concerns since it gives the impression of representing a cutting edge scientific development, with considerable therapeutic potential. However, regenerative medicine poses distinct problems for many commentators mainly because of the involvement of stem cells and the far-reaching and frequently unjustified

---

1   L. R. Kaiser, 'The Future of Multihospital Systems', *Topics in Health Care Financing* 18/4 (1992), 32–45.
2   Thomson et al., 'Embryonic Stem Cell Lines Derived from Human Blastocysts'.
3   King, Whitaker, and Jones, 'Speculative Ethics'.
4   V. Mironov, R. P. Visconti, and R. R. Markwald, 'What Is Regenerative Medicine? Emergence of Applied Stem Cell and Developmental Biology', *Expert Opinion on Biological Therapy* 4/6 (2004), 773–81.

hopes emanating from them.[5] These hopes and consequent fears lie not in the therapeutic potential of stem cells but in their postulated role in some radical redesign of human nature.[6] In this way regenerative medicine has come to be depicted as being 'rich with Promethean promises'.[7] This represents a move from the growth of new tissues and organs in the laboratory, and with this the prospect of changing the course of chronic disease, as first envisaged by Kaiser,[8] to the idealistic vision of conquering disease and opening the way to ever lengthening life spans. Once the latter course is embarked upon, the realm of transhumanism has been ushered in, and with it an exacerbation of social inequalities, intergenerational fairness, environmental ethics, and the problems posed by endless life spans, with subsequent divergence of enhanced and unenhanced human species.[9]

The confluence of the notion of regenerative medicine and attempts at enhancing human capabilities and life span comes out clearly in the writings of some commentators.[10] And so, acceleration in the healing of skin wounds may, it is envisaged, lead to the elimination of wrinkles in the facial skin of an octogenarian. Taken even further, it is feared that regenerative medicine will be devoted to realizing its potential to prolong life. Such fears are grounded by reference to the expansive vistas painted by Silver and Stock based on combining germ-line genetic engineering and

5   C. J. Kirkpatrick et al., 'Visions for Regenerative Medicine: Interface between Scientific Fact and Science Fiction', *Artificial Organs* 30/10 (2006), 822–7.

6   Nick Bostrom, 'A History of Transhumanist Thought', *Journal of Evolution and Technology* 14/1 (2005), 1–25; Walter Glannon, 'Decelerating and Arresting Human Aging', *Medical Enhancement and Posthumanity* (2008), 175–89; King-Tak Ip, ed. *The Bioethics of Regenerative Medicine* (Netherlands: Springer, 2009).

7   King-Tak Ip, 'Introduction: Regenerative Medicine at the Heart of the Culture Wars', in King-Tak Ip, ed. *The Bioethics of Regenerative Medicine* (Netherlands: Springer, 2009), 3–10, 3.

8   Kaiser, 'The Future of Multihospital Systems'.

9   Nicholas Agar, 'Whereto Transhumanism? The Literature Reaches a Critical Mass', *Hastings Center Report* 37/3 (2007), 12–17.

10   Gerald P. McKenny, 'The Ethics of Regenerative Medicine: Beyond Humanism and Posthumanism', in King-Tak Ip, ed. *The Bioethics of Regenerative Medicine* (Netherlands: Springer, 2009), 155–69.

reproductive technologies.[11] Regenerative medicine is viewed as integral to these grand vistas, with its potential for enhancing human abilities.[12] While I have considerable sympathy with the critique of these vistas, the role of regenerative medicine in bringing them about is far from convincing. The result is that regenerative medicine has been given a role in posthuman aspirations, a role that I regard as highly questionable.

Ethical debate on regenerative medicine has also become embroiled in the stem cell debate, and in particular on the use of embryonic stem cells.[13] All the vitriol surrounding use of these cells has emerged in the regenerative medicine debate, even though the span of the latter extends far beyond embryonic stem cells. For instance, tissue engineering to produce a replacement airway (bronchus) for a patient with loss of their normal airway utilized a donor trachea plus epithelial cells and mesenchymal stem cell-derived chondrocytes from cells taken from the recipient.[14]

Preoccupation with the moral status of embryos, while a legitimate ethical debate, is far from central to the regenerative medicine debate. Lysaght and Campbell have argued that bioethicists must give due attention to the largely neglected issues of informed consent processes, the exploitation of women, the commodification of human tissue, science communication and the ownership of immortal cell lines.[15] All these are core ethical considerations for regenerative medicine as it seeks to enter the clinic. Speculative scenarios, with little if any relation to current clinical practice, such as the remaking of human nature through regenerative medicine, threaten to direct attention away from pressing current and emerging issues.

11   L. M. Silver, *Remaking Eden: Cloning and Beyond in a Brave New World* (New York, NY: Aven Books, 1997); Stock, *Redesigning Humans*.

12   McKenny, 'The Ethics of Regenerative Medicine'.

13   Wyatt, *Matters of Life and Death*.

14   P. Macchiarini et al., 'Clinical Transplantation of a Tissue-Engineered Airway', *Lancet* 372/9655 (2008), 2023–30.

15   T. Lysaght and A. V. Campbell, 'The Ethics of Regenerative Medicine: Broadening the Scope Beyond the Moral Status of Embryos', in Akira Akabayashi, ed. *The Future of Bioethics: International Dialogues* (Oxford: Oxford University Press, 2014), 5–26.

Once again it is imperative to differentiate between hype and realism, between speculation and what can be done to assist patients in the foreseeable future. This is crucial since without this the efforts of serious researchers may be hampered by unhelpful media distortion and political ineptitude. Theologians may also devote far too much attention to combating starry eyed humanistic ideals rather than seeking ways in which those caring for patients can be better equipped to improve the quality of the lives of their patients and therefore their prospects of flourishing as much as possible.

In spite of the serious clinical prospects held out by regenerative medicine, the field faces complex difficulties that are hampering the clinical application of stem cells at the most basic level. For example, scientists are yet to ascertain how to reliably direct cell differentiation to the desired lineage and modify cells without raising the risk of tumour formation.[16] It is not anticipated that solutions to these basic problems will come quickly; as with most scientific advances in complex fields, developments tend to come gradually so that hype should always be tempered with caution.[17]

King et al have commented that any discussion that uncritically conflates regenerative medicine and its likely prospects with grandiose claims about remaking what it means to be human is profoundly unhelpful.[18] This may result in regenerative medicine being tainted by condemnation of vast claims that are only rarely made by the scientists intimately involved in the laboratory and clinical research. This is unlikely to be the intention of the commentators who are critical of humanistic claims, but they are obligated to distinguish carefully between the speculative hype and the serious science. The goal of regenerative medicine is not to view the human body as infinitely plastic; this is self-evidently not the case. It is even further from the truth to accuse these scientists of denying human finitude and mortality, and yet this is the concern of some who are deeply

16 Kirkpatrick et al., 'Visions for Regenerative Medicine'.
17 G. Q. Daley, 'Stem Cells: Roadmap to the Clinic', *The Journal of Clinical Investigation* 120/1 (2010), 8–10.
18 King, Whitaker, and Jones, 'Speculative Ethics'. See, for example, Ip, ed. *The Bioethics of Regenerative Medicine.*

concerned by the speculative hype.[19] It is those with transhumanist pretensions who indulge in this speculation, not those committed to serious scientific endeavours.

In chapter 3, I dealt with the ARTs and with the crucial role of IVF. There it emerged that Christian theologians espousing a wide variety of theological positions have expressed their deep unease at the intrusion of technology into what has traditionally been regarded as profoundly human territory. One may or may not consider it peculiarly 'divine' territory but it has characteristics that seem to many to set it apart from other areas where the intrusion of medicine is generally welcomed.

While the ARTs do not fall under the umbrella of regenerative medicine, I want to raise one aspect that has comparable connotations. This concerns attempts to create fertile eggs from stem cells, and subsequently breed healthy offspring from the eggs so created. In mice, stem cells (embryonic stem cells or induced pluripotent stem cells) have been genetically reprogrammed to become egg precursor cells. When mixed with selected body cells from female mice, they produced 'reconstituted ovaries', and when these were implanted into mice they developed into mature eggs. The next step was the fertilization of these eggs by IVF and the subsequent birth of mouse pups.[20] While the process at present is very inefficient, it marks a major milestone in understanding how eggs are produced raising the prospect that one day it may prove feasible to stimulate the production of eggs for infertile women. The same research group has also created viable sperm from mouse stem cells.[21]

There is no hint even here of being driven by images of creating 'perfect babies'. The thrust is consistently that of helping people unable to conceive.

---

19   Robert Song, 'Genetic Manipulation and the Resurreciton Body', in King-Tak Ip, ed. *The Bioethics of Regenerative Medicine* (Netherlands: Springer, 2009), 27–45.

20   K. Hayashi et al., 'Offspring from Oocytes Derived from In Vitro Primordial Germ Cell-Like Cells in Mice', *Science* 338/6109 (2012), 971–5; A. Jha, 'Baby Mice Created from Stem Cells', *The Guardian* (4 October 2012) <http://www.guardian.co.uk/science/2012/oct/04/baby-mice-stem-cells> accessed 24 June 2013.

21   K. Hayashi et al., 'Reconstitution of the Mouse Germ Cell Specification Pathway in Culture by Pluripotent Stem Cells', *Cell* 146/4 (2011), 519–32.

Even the use of stem cells to create cardiac cells from general tissues like fibroblasts or skin has a therapeutic rationale, even though the spectre that regenerative medicine will one day be used to remake humans in some new image is raised from time to time.[22]

Cyborgs

While regenerative medicine can mean different things to different people, the general impression is that, at least in its immediate manifestation, it has therapeutic overtones. After all, it is 'medicine', even if a highly sophisticated version of medicine. Cyborgs, however, are far more forbidding, with their overtones of the organic and the synthetic, the natural and the artificial, that leads to visions of part human – part machine. While this entails a continuum from standard medical prostheses through to a host of examples in the military and in science fiction, it is the latter that have captured the public imagination.

The existence of prostheses elicits no particular surprise, since prosthetic limbs and hands, and prosthetic joints are de rigueur. No longer are they the stuff of science fiction. Imperceptibly, as the sophistication of the procedures has increased, they have begun to incorporate elements usually regarded as artificial. They are still little more than a routine aspect of modern medicine, and in no way are they viewed as threatening one's sense of self-identity. After all, what is the essential difference between an artificial limb and a crutch? The former is far more efficient and gives the appearance of (semi-) permanence, and this will generally be welcomed. Nevertheless, it contains within itself elements of a human-machine interface, however rudimentary these may be, and it is these that are beginning to transform its

22   C. Mummery et al., 'Differentiation of Human Embryonic Stem Cells to Cardiomyocytes: Role of Coculture with Visceral Endoderm-Like Cells', *Circulation* 107/21 (2003), 2733–40.

nature. This is taken much further when the limb, particularly an arm and hand, are brought under control of the individual's own brain.[23] Although this interface is far more intrusive, it is simply replicating normal events in the human body. The relevant part of the brain is being brought into play to control what is essentially a prosthetic device (as discussed in chapter 5).

To a degree we are already living in a prosthetic world, in the sense that we are relying increasingly on technology, both external to us (smartphones) and within our bodies (artificial joints). Is this simply the beginning of a continuing trend in this direction, and why are we acting like this? Is it a desire to transcend our humanness, on the ground that the limitations inherent within normal human living can and should be overcome? Or is it the hope that this will improve our functioning as human beings? Brain implants in their various forms have been discussed in chapter 5, leaving room for consideration of more extreme mergers of human and machine. With the sudden explosion in machine intelligence and rapid innovations in gene research and nanotechnology, the distinction between biological and mechanical, and physical and virtual reality is becoming blurred.[24]

Ray Kurzweil in his book, *The Singularity is Near*,[25] foresees a theoretical future period of extremely rapid technological progress when accelerating technology will lead to superhuman machine intelligence that will soon exceed human intelligence. In his view, human existence on this planet will be irreversibly altered. This will be due to a combination of brain power with computer power. The effect of this, he envisages, will be a transformation of the knowledge, skills and personality quirks that make us human, enabling us to think, reason, communicate and create in ways unimaginable today. The distinction between the biological and the mechanical, and between physical and virtual reality will be obliterated. The opportunities opened up by these developments will, in his words, allow us to transcend our frail bodies with all their limitations. Illness, as

---

23   R. Kwok, 'Neuroprosthetics: Once More, with Feeling', *Nature* 497/7448 (2013), 176–8; Velliste et al., 'Cortical Control of a Prosthetic Arm for Self-Feeding'.
24   Jones and Whitaker, 'Transforming the Human Body'.
25   R. Kurzweil, *The Singularity Is Near* (New York, NY: Viking, 2005).

we know it, will be eradicated. Through the use of nanotechnology, we will be able to manufacture almost any physical product upon demand, world hunger and poverty will be solved, and pollution will vanish. Human existence will undergo a quantum leap in evolution. We will be able to live as long as we choose.[26]

Kurzweil describes what he calls the 'Human Body 2.0'.[27] A modified, up-dated human body where nanobots, robots the size of blood cells, have provided the means to conceptually redesign every body system. He continues:

> We are becoming Cyborgs. We are rapidly growing more intimate with our technology. Computers started out as large remote machines in air-conditioned rooms tended by white-coated technicians. Subsequently they moved onto our desks, then under our arms, and now in our pockets. Soon, we'll routinely put them inside our bodies and brains. Ultimately we will become more nonbiological than biological.[28]

Eventually he sees the line between humans and machines blurring as artificial intelligence develops, and humans embrace cybernetic implants. Kurzweil's projections may be far removed from what will actually eventuate, but the genesis of his imagined world is current reality. Of course, the vastly expanded trajectory he imagines is far from inevitable. The world predicted by Kurzweil may prove illusory, as may the more recent '2045 Initiative' with its goal of producing lifelike avatars uploaded with the contents of a human brain and complete with consciousness and personality.[29] But what if some such cyborgs do eventuate? What then? To some degree it all depends what one means by cyborg, since there is a continuum from

---

26 R. Kurzweil, 'Reinventing Humanity: The Future of Human-Machine Intelligence', *The Futurist* 40/2 (2006), 39–46.

27 R. Kurzweil, 'Human Body Version 2.0', (16 February 2003) <http://www.kurzweilai. net/human-body-version-20> accessed 11 April 2013.

28 Kurzweil, 'Human Body Version 2.0'.

29 David Segal, 'This Man Is Not a Cyborg. Yet', *New York Times* (1 June 2013) <http:// www.nytimes.com/2013/06/02/business/dmitry-itskov-and-the-avatar-quest.html?_ r=0&adxnnl=1&pagewanted=all&adxnnlx=1372069074-ckcc4DUQvoAkkAIrx- BOoSw> accessed 24 June 2013.

humans with a single prosthesis (an arm or leg or heart valve) to those in a hypothetical future more machine than human.

Andy Clark proposes that we are natural-born cyborgs, a concept derived from the observation that the drive to incorporate technology is so much a part of our human nature that technology cannot be separated from our essential selves.[30] For him, enhancements like language, written text, and digital encodings are integral parts of what we are as cognitive beings. In his view these enhancements are a natural extension of our human nature, and the integration of technology will not stop at external aids. He writes, 'Human-machine symbiosis, I believe, is simply what comes naturally. It lies on a direct continuum with clothes, cooking ("external artificial digestion"), bricklaying, and writing. The capacity to creatively distribute labor between biology and designed environment is the very signature of our species'.[31]

Our partnership with computers today is far more pervasive than most realize, since our use of them leaves data shadows; they 'remember' us in a way in which a photocopier does not. This can be, and has been, commercially exploited; the privacy concerns raised by the increasing mingling of our computer use and personal identity are significant.[32] In other words, even now our partnership with computers is a symbiotic one.[33] Another manifestation of this development are the widely expressed concerns over data surveillance programs, and their potential to be used to unknowingly obtain the private information of individuals and even whole societies.

I shall argue that 'true cyborgs' will be those beings where the interface directly intrudes into the brain, and when the intrusion significantly alters the person. Gillett refers to 'a non-human mode of relationship and

---

30 Andy Clark, *Natural-Born Cyborgs: Minds, Technologies, and the Future of Human Intelligence* (Oxford: Oxford University Press, 2003).

31 Clark, *Natural-Born Cyborgs*, 174.

32 Clark, *Natural-Born Cyborgs*, 169–74.

33 Board for Social Responsibility of the Church of England, *Cybernauts Awake!* (London: Church House Publishing, 1999).

reaction or response to others'.[34] Hence, it may be wise to refer to cyborgs only when the intrusion of the artificial into the human results in significant functional differences cognitively. This will probably involve implants in decision-making brain regions. It is within this context that we encounter concerns that cyborgs will disrupt normal moral categories,[35] although one imagines that the machine element would have to be a dominant one before anything of this nature emerges. At some point it will become important to establish criteria for determining how much artificiality can be incorporated into a subject before they are transformed from human into cyborg.[36]

As long as computers remain external to the human body, society appears to be largely unconcerned about the implications, even though as external appendages they have enormous implications for what we are as people. Comparing computers to artificial limbs is instructive. It is when technology is internalized that negative responses come to the surface, as though the artificial is intruding into what we are as humans. This is the interface of the natural and artificial, the biological and cybernetic. Even before computers are fully internalized unsettling degrees of assimilation are encountered. MIT's wearable computer project envisions a future where computers are a clothing accessory, almost seamlessly augmenting daily living.[37] The possibilities opened up by cybernetic transplants take us to the limits of our imaginations.

For some the image of cyborg is one to be resisted, on the ground that the intrusion of too much that is artificial in the brain would probably result in a creature lacking in truly human capacities. For Christian writers moves in a cyborgian direction represent a human challenge to God's status as the one and only creator. This is of course speculation since this

---

34   G. R. Gillett, 'Cyborgs and Moral Identity', *Journal of Medical Ethics* 32/2 (2006), 79–83.

35   Donna Haraway, *Simians, Cyborgs, and Women: The Reinvention of Nature* (London: Free Association Books, 1991), 149–55.

36   Gillett, 'Cyborgs and Moral Identity'.

37   MIT Media Lab, 'Wearable Computing at the MIT Media Lab', (2005) <http://www.media.mit.edu/wearables/> accessed 11 April 2013.

experiment has never been undertaken, and may never be scientifically feasible. Consequently, discussions of what may be termed 'radical cyborgs' will it seems remain in the realm of the speculative.

A radical cyborgian future?

In contrast to the commentators previously discussed there are voices that see a far more positive future for cyborgs, and view them as having a positive role in a human future. One of these is Haraway, who challenges clearly defined distinctions between the natural and artificial, and who advocates a blurring of the boundary between organism and machine.[38] For her, human identity is not fixed and so it is misleading to reject cyborg visions on the basis of absolutist categories. According to Deane-Drummond, Haraway seeks to deconstruct the 'ontological hygiene' that underlies Western modernity's classifications systems through class, race, gender or species.[39] For Haraway, the machine is an extension of who we are so that 'The machine is us, our processes, an aspect of our embodiment.'[40] According to Graham, Haraway's approach has strong overtones of transhumanist thinking.[41] However, Deane-Drummond doubts this, since in her view transhumanists have the specific goal that human life becomes identified with particular rarified mental capacities.[42]

38   Deane-Drummond, 'Bodies in Glass'; Donna Haraway, 'Cyborg to Companion Species: Reconfiguring Kinship in Technoscience', in Donna Haraway, ed. *The Haraway Reader* (London: Routledge, 2004), 295–320.
39   Deane-Drummond, 'Bodies in Glass'. For an excellent commentary on Haraway's approach to cybernetics, see Elaine L. Graham, *Representations of the Post/Human: Monsters, Aliens and Others in Popular Culture* (Manchester: Manchester University Press, 2002), 200–12.
40   Donna Haraway, 'A Cyborg Manifesto: Science, Technology and Socialist Feminism in the Late Twentieth Century', in Donna Haraway, ed. *Simeans, Cyborgs and Women* (London: Free Association Books, 1991), 149–81, 180.
41   Graham, *Representations of the Post/Human*, 204.
42   Deane-Drummond, 'Bodies in Glass'.

While these transhumanist vistas seem far removed from the world of cyborgs, they could incorporate radical cyborgs into their grand projections of where human society might go. Intensely speculative as all this is, the practical application is that it opens the way to greater acceptance of the far more routine uses of increasingly intrusive prostheses, neural implants and genetic manipulations. For Deane-Drummond, the transhuman trajectory illustrates the folly of the mechanistic application of the artificial in an attempt to solve human limitations and problems, characterized as it is by societal injustices, and the collapse of the relational aspects of human existence.[43] In another context Deane-Drummond concludes: 'Perhaps we have created the cyborg from our own imaginings, and it is time to give it the qualified place that it needs, namely, as a creation of human ingenuity but one that needs to be firmly kept in its place: technology needs to remain our servant, not our master or our goal'.[44]

Far away from the highly speculative, many of the aspects of cyborgs reflect surprisingly traditional ethical questions. Before attempting to move in these directions, it is imperative to reflect on whether developments will impair human self-identity, whether there is the prospect of implementing them voluntarily or whether coercion would be involved. Will they benefit people at an individual level and what might be the effects on society? Ongoing debate on the prospects opened up by cyborgs needs to face up to these mundane, but crucial, considerations.

Writing in an expansive evolutionary context Hefner views cyborgs as occupying a place in a future evolutionary trajectory, bestowing this with religious aspirations.[45] For him, 'Cyborg is created in the image of God [...] Technology is now a phase of evolution, and it is a new creation, a vessel for the image of God'.[46] According to him not only is God involved in technological developments, but his goals are reflected in those of humans, such

---

43 Deane-Drummond, 'Bodies in Glass'.

44 Celia Deane-Drummond, *Christ and Evolution: Wonder and Wisdom* (Minneapolis, MN: Fortress Press, 2009), 285.

45 Philip Hefner, *Technology and Human Becoming* (Minneapolis, MN: Fortress Press, 2003), 43.

46 Hefner, *Technology and Human Becoming*, 77.

that in human-created cyborgs, we can appreciate more of the work and aspirations of God. This presupposes that all technological developments undertaken by humans are appropriate and have good ends in view, regardless of ethical let alone theological criteria. Each technological venture has to be assessed and critiqued on its own merits. Divorced from such analysis, one ends up with a naïve view of the progression of technology. Beyond this it is arrogant to claim that any particular direction in which technology is taken is in line with God's imperatives. Deane-Drummond comments: 'Whether or not the cyborg is the image of God is not open to human inquiry, for the only image that we have been given [...] is Christ, who is the manifestation of God and humanity'.[47]

Cyborgs as human creations will always remain precisely this: created by humans to benefit humans. Within a Christian perspective they are tools to serve human beings who alone are the image of God. In view of this, if cyborgian artefacts begin to transform humans in ways that give the appearance of dominating human existence, they are to be rejected and research along such lines is to be rejected. In my view it is highly unlikely that this will ever become reality.

Post-persons

The fringes of bioethical debate are taken up with discussions of post-persons and their role within some of the vast agendas of transhumanism.[48] Reactions to speculative theorizing of this ilk range from fascination to disinterest, and from enthusiasm to perplexity. Some, including myself, wonder why we should be discussing the moral implications of modified humans who do not as yet exist and may never exist. Is this really where technology is leading? The answer may be straightforward: this is not

47   Deane-Drummond, *Christ and Evolution*, 265.
48   Garreau, *Radical Evolution*.

where it is leading any more than this is where cyborgs are heading. And yet, what if some of these developments do emerge, even in small part? What implications might that have for Christian thinking and attitudes?

An Italian politician, Giuseppe Vatinno, considers that transhumanism aims to free humanity from its biological limitations, overcoming natural evolution and thereby making us more than human.[49] Even if it makes us less than human, that too may be a good thing because it could mean the end result may be less subject to the whims of nature, including illness or extremes of climate. Vatinno describes himself as an 'extropian', a species of transhumanist, which strives for practical improvements in human welfare, like life extension technologies, research into artificial intelligence, space exploration, nanotechnology and biotechnology.

According to Vatinno, these directions stem from science and the scientific method that in turn usher in ethical values based upon the absolute honesty and search for the truth embedded within science. They could serve as a model of correctness. These values have religious connotations, even if of a non-supernatural variety. At first glance it seems strange that science is entering the realm of religion, since the scientists concerned are not religiously inclined. Once again, this gives the appearance of being *Homo sapiens* attempting to achieve salvation by transcending itself through technology, on the assumption that this is possible or at least will become possible within a short time span.

Outlandish as some of these prospects appear, they have entered the arena of serious academic debate. Consider an academic book entitled *Building Better Humans?* with the subtitle *Refocusing the debate on transhumanism.*[50] It is described as 'the first comprehensive, multi-disciplinary, inter-religious, and critical engagement with transhumanism as a cultural phenomenon, an ideology and a philosophy. Situating transhumanism in its proper historical context, the essays reflect on transhumanism from the

49 Edwin Cartlidge, 'Meet the World's First Transhumanist Politician', *New Scientist* (18 September 2012) <http://www.newscientist.com/article/mg21528826.100-meet-the-worlds-first-transhumanist-politician.html > accessed 12 April 2013.

50 H. Tirosh-Samuelson and K. L. Mossman, eds, *Building Better Humans? Refocusing the Debate on Transhumanism* (Frankfurt am Main: Peter Lang, 2012).

perspectives of several world religions, ponder the feasibility of regulating human enhancement, tease out the philosophical implications of transhumanism, explore the interplay between technology and culture, and expose the scientific limits of transhumanism'.

Expansive scenarios are everywhere within transhumanist thinking, and hence a book along these lines is entirely appropriate. The usual limits experienced by human beings will have disappeared. Not only will the major diseases that afflict present-day societies be banished, from diabetes and heart disease, to Parkinson's and Alzheimer's diseases, but the regeneration of all tissues and organs will be readily accomplished. The end result will be almost inevitable; human immortality will beckon as a physical phenomenon.[51] Death will have been vanquished; all will be made new through the efforts of human beings and their technological achievements (see chapter 7). This is indeed the world of post-persons.[52]

Discussions around post-persons have been stimulated by the thought that it may prove feasible to enhance cognitive and moral capacities to a level beyond that possessed by mere persons, that is, existing persons.[53] If this does eventuate would it be morally wrong to strive to bring about this new form of being? Buchanan argues that moral status would not be affected, even if some individuals were enhanced to become more rational and intuitive than others. Agar, however, considers that post-persons would have a higher moral status and should not be brought into existence.[54] His reasoning is that we have no moral justification to bring post-persons into existence, and that were we to do so there may well be tragic consequences for mere persons. For instance, post-persons would be morally justified in sacrificing the lives of mere persons for their own ends, and this is ethically

51   Jones, 'Enhancement: Is Baseless Speculation Misleading Theologians and Bioethicists?'.
52   Nicholas Agar, 'Why Is It Possible to Enhance Moral Status and Why Doing So Is Wrong?', *Journal of Medical Ethics* 39/2 (2013), 67–74; Michael Hauskeller, 'The Moral Status of Post-Persons', *Journal of Medical Ethics* 39/2 (2013), 76–7.
53   Allen Buchanan, 'Enhancement and the Ethics of Development', *Kennedy Institute of Ethics Journal* 18 (2008), 1–34.
54   Agar, 'Why Is It Possible to Enhance Moral Status and Why Doing So Is Wrong?'.

unacceptable. While the details of this debate are far removed from my immediate concerns it has elicited spirited debate, revolving around the nature of the arguments put forward by Agar.[55]

One of the more interesting commentaries on Agar's position is that of Sparrow, who while in agreement with his conclusions, is more apprehensive about the proclivity of post-persons to harm (mere) persons.[56] His reason is that currently human persons abuse sentient non-human animals and exploit members of the animal kingdom. Whether or not one agrees with all his concerns, he is correct in drawing attention to the readiness with which humans treat even other humans (ostensibly those with equal moral status) with disdain and sometimes with appalling cruelty. He also observes the willingness with which some philosophers actively pursue their own obsolescence by advocating for the creation of those with a higher moral status than they possess!

Sparrow touches on a pertinent point frequently overlooked in discussions on moral enhancement, and particularly in reference to post-persons.[57] This is the reality of evil that blights our present existence and is hardly likely to be swept away by technological manipulations. Why should we expect post-persons to always act in self-effacing and altruistic ways? There is no evidence that they would act in these ways, and by the nature of what is being proposed there can be no way in which evidence could be obtained prior to their emergence. Experience with mere persons is disheartening in this regard, and so a cautious approach would be to dissuade anyone from moving in this direction. Post-persons may even be more aggressive and self-centered than any mere persons are now. We simply do not know, and we cannot find out without undertaking an unethical experiment.

55   R. Powell, 'The Biomedical Enhancement of Moral Status', *Journal of Medical Ethics* 39/2 (2013), 65–6.

56   R. J. Sparrow, 'The Perils of Post-Persons', *Journal of Medical Ethics* 39/2 (2013), 80–1.

57   Sparrow, 'The Perils of Post-Persons'.

The same argument was made against IVF prior to its introduction, and yet there is a difference.[58] IVF was introduced with the aim of overcoming infertility; it had ethical justification, even though this reasoning did not persuade everyone. I can find no equivalent justification for morally enhancing people with the aim of producing post-persons. Even while agreeing that moral improvement is urgently required in most societies, a technological means of achieving this is an ideological claim and not one based on convincing scientific evidence. This is a questionable basis for the further move towards creating a population of post-persons. It would be better to assess how moral behaviour can be improved in mere persons, and this is where Christian contributions enter the picture.

Theological challenges

In this chapter I have presented three models of intrusion into the human body and therefore the human condition: regenerative medicine with its many therapeutic possibilities, the production of cyborgs with their increasing reliance upon non-biological interventions in the brain and body, and post-persons with their postulated increased moral status. These three examples represent an approximate continuum from cutting edge biotechnology, to an extension to current ways of utilizing non-biological means of assisting bodily function, and on to speculative future possibilities of transforming human beings as they are currently known and understood. All three are dependent upon biomedical technologies, and all three incorporate procedures that may prove valuable to those suffering from illnesses and infirmities. The extent to which the latter play a role in all three varies though, as does the realism of the proposals. The balance

58   Paul Ramsey, 'Shall We "Reproduce"? I. The Medical Ethics of in Vitro Fertilization', *JAMA* 220/10 (1972), 1346–50; Paul Ramsey, 'Shall We "Reproduce"? II. Rejoinders and Future Forecast', *JAMA* 220/11 (1972), 1480–5.

between a therapeutic imperative and a technological imperative varies considerably across this continuum.

The challenge is to attain a balance that keeps the therapeutic imperative uppermost, and that is prepared to critique both the extreme visions, whether of utopia or dystopia. Notions of progress and rationality, as well as of compassion and hope, can be used to advocate for all manner of technological developments. Against this, doomsday scenarios serve to frighten and scare with their visions of modified humans with altered life courses, enshrining cyborgian elements and lacking human warmth and relationships. Each of these in its different ways is misleading, and we have to learn to reject both paths. We have to look for a far more moderate and realistic response, searching for that which is helpful and avoiding the hype and extremism. This I believe is what we should be able to find in Christian understanding.[59]

Before I move into this detail however, I should take a step back and engage with a theological position with which I have concerns. There is a strand of very serious theological argument that dismisses many of the technological developments with which I have been dealing on the ground that it detracts from dependence upon God and from the givenness of his creation. This in turn sets up an apparent trade-off between a biblical/religious stance and a scientific/technological one. This is thoroughly worked out by Northcott in his analysis of cloning, and his critique of 'the theology of making'.[60] In discussing cloning, he writes, 'Principled religious opposition [...] arises from the sense that with this technique humans have begun to play with the very constitution of life itself, and that this takes human creativity beyond the boundaries of morality and wisdom'.[61] His opposition stems from the human act of creation and the human invention

---

59   D. Gareth Jones, 'The Importance of Realism in Assessing Technological Possibilties: The Role of Christian Thinking', *Christian Perspectives on Science and Technology, ISCAST Online Journal* 9 (June 2013) <http://www.iscast.org/journal/articles/Jones_G_2013-06_Realism> accessed 24 June 2013.

60   M. S. Northcott, 'Concept Art, Clones, and Co-Creators: The Theology of Making', *Modern Theology* 21/2 (2005), 219–36.

61   Northcott, 'Concept Art, Clones, and Co-Creators', 224.

of life, which point he claims to human self-transcendence and the desire to reorder 'the constitution of his own embodied mammalian being.'[62] Northcott has grave doubts about the scientific imposition of order on matter and about the manipulation of the elements of life. For him such moves indicate 'radical disrespect, distrust and denial of any intrinsic beauty, goodness or truth in the original ordering of life itself'.[63] As in other areas I would point to the hyperbole in these statements, not to deny the fundamental issues at stake, but what appears to be wholesale rejection of the scientific enterprise. This may not be the intention, since it is unlikely that scientific contributions to an understanding of the cellular mechanisms underlying infectious diseases, cancers, developmental abnormalities and neurological dysfunction, will be rejected. And yet all these interfere with God's creation and the ordering of life. Theological critiques of scientific intrusions into the human body need to be more nuanced, otherwise it will be all too easy to construe theological engagement with science as being driven by a negative agenda.

Perfection and self-perfection

To varying degrees, there is a longing for self-perfection in each of the realms I have touched on. Some Christian commentators relentlessly attack this as espousing a God-denying culture, driven by the dream of human perfection and intent upon enhancing natural capabilities.[64] One commentator has written: 'a mortality-denying, imperfection-denying culture that is implicit in much of the modern project will not be able to accept mortality and

---

62   Northcott, 'Concept Art, Clones, and Co-Creators', 229.
63   Northcott, 'Concept Art, Clones, and Co-Creators', 230.
64   Celia Deane-Drummond, 'Future Perfect? God, the Transhuman Future and the Quest for Immortality', in C. Deane-Drummond and P. Scott, eds, *Future Perfect? God, Medicine and Human Identity* (London: T&T Clark, 2006), 168–82; Junker-Kenny, 'Genetic Perfection'.

imperfection as such, expressed paradoxically through insidious policies of eugenics and euthanasia.'[65]

I have some sympathy with the gist of this condemnation, but I do not consider that it applies equally to the three realms (regenerative medicine, cyborgs, post-persons). Nevertheless, it serves to remind us that for Christians ultimate perfection is to be found in God alone and in his redeemed kingdom. And so, while it is true that we are to strive to be perfect and to be holy, we do not expect to attain this state in the present kingdom. Additionally, the perfection we are to seek is to be that of character and attitudes and not of the physical human body. It is within the network of human and divine relationships that we put into effect the work of Christ. This transcends the physical and biological, but neither does it totally ignore them. Hence I would hesitate to condemn genetic intrusions into embryos (if it can be shown that these will accomplish what they set out to accomplish to benefit the future individuals), since the goal of these is not genetic perfection. They fit within the paradigm of healing and care for those in need of restoration, albeit restoration that is limited in both time and scope. Such goals are realistic, and fit comfortably within Christian imperatives.

Alongside talk of perfection, it is important to be reminded of our imperfection as human beings. Everything we touch is tainted; we see in a glass darkly (1 Cor. 13:12). We never see completely. Our understanding is inevitably partial, and we are never as wise as we wish to be. Consequently, our scientific endeavours and our clinical competence are incomplete; the developments of which we are most proud leave much to be desired and Christians should be the first to applaud what we can do but also acknowledge what is beyond our powers of comprehension and control. Self-perfection is unattainable biologically and untenable theologically.

---

65   Celia Deane-Drummond, 'How Might a Virtue Ethic Frame Debates in Human Genetics?', in Celia Deane-Drummond, ed. *Brave New World: Theology, Ethics and the Human Genome* (London: T&T Clark International, 2003), 225–52.

Mortality and immortality

The more extreme versions of the technological aspirations constitute a secular eschatology, in which humans will be able to achieve a form of bodily immortality. This is where transhumanism enters the picture, since this is its explicit aim. In this aspiration the future becomes an extension of the present, with all the foibles and problems of the present. Hope resides in this continuation, and in nothing else. However, the idealism inherent within this perspective contends that the foibles and problems can be removed by technology. In other words, continuation of the present is made possible by removing all pathologies that lead to illness and ageing using the powers of technology, thereby ushering in perfection and immortality. Any future existence is a vastly improved version of the present life.[66] This is precisely the tenor of vision of which Christian writers like Deane-Drummond and Junker-Kenny are so critical.[67]

Apart from the superficiality of this vista, Christians would want to point out the lack of any recognition of human sinfulness, with its incursion into all human activities and aspirations. They would also want to remind proponents of these optimistic scenarios of the ever-present reality of death and loss. It is all very well postulating a world without death, but there are not even suggestive hints of this in any society today.

Facing up to the reality of death brings us to the heart of Christian thinking. Christians should not extol the virtues of death since death is real and is an evil. Verhey writes, 'Death sunders human beings from their own flesh, from the community of praise, and from God. Death is a power that threatens [...]'.[68] It threatens an unraveling of meaning and is always a cause of sorrow and grief, but the context for the Christian is one of hope based in the power of God that raised Jesus from the dead.[69] Consequently, we are not to seek our hope in technological mastery over nature, 'but rather

---

66   Garreau, *Radical Evolution*.
67   Deane-Drummond, 'Future Perfect?'; Junker-Kenny, 'Genetic Perfection'.
68   Verhey, *The Christian Art of Dying*, 191.
69   Verhey, *The Christian Art of Dying*, 201.

in the creative work of God that can call a cosmos out of chaos and give light to the darkness and life to the dust'.[70] Since technology is not to be the basis of our hope, with its presumption and ultimate despair, Christians are freed to care for others even while they lie dying. In recognizing the 'not yet' character of our present existence, we can come to terms with mortality, both our own mortality and that of others.

The realism of a Christian diagnosis forces us to take note of the inequalities of opportunity throughout the world, where speculation about endless biological life must be little more than a sad joke to the majority of people alive today. One commentator has written: 'Such drives avoid facing the tragic reality of a life cut off well before its prime, and the added injustices associated with uneven distribution of medical resources that make consideration of life extension and other enhancements the privilege of a relatively small minority, even if desired more widely'.[71] Caring for people in dire need, treating eminently treatable diseases and raising the standard of living of countless people constitute the only form of enhancement consonant with Christian aspirations. By accepting our mortality, we can devote ourselves to serving others as they face suffering and deprivation, within a context in which immortality revolves around the purposes of a good God who suffers with us because he loves and cares for all.

This, in turn, leads to commitment to all who are impaired, including those who are genetically impaired. When divorced from aspirations towards genetic perfection, we are freed to do all we can to alleviate the plight of those whose genetic constitution precipitates serious disease. This falls into the category of health-related enhancements that are far removed from desire-fuelled enhancements.

Underlying Christian expectations is the Christ-centered hope that God will bring into being a world redeemed and redirected. It is the hope of the resurrection and of resurrected bodies in which all are made new.[72] The shackles of the old will have been thrown off, and the new creation

---

70   Verhey, *The Christian Art of Dying*, 263.
71   Deane-Drummond, 'Future Perfect?', 176.
72   Wright, *Surprised by Hope*.

will be inaugurated. However, this new creation is not a natural extension of the present sinful world, and will not be brought about by human technology or human initiative. While neither of the latter is to be shunned, their effects are limited. Christians are to be grateful for what can and has been achieved technologically but this will never usher in the new heavens and the new earth.

Finiteness and limitations of created humans

The discussion so far in this chapter has barely touched on the prospects opened up by moral enhancement. One could argue that Christians should welcome any form of enhancement, and so it seems churlish to question prospects presented by moral enhancement, even if dependent upon technological intervention. My concerns from a Christian angle will have come through in chapter 6. To expect too much of technology in this area overlooks the finiteness of human endeavours. This is not reason to reject outright attempts to improve morality, but one gets the impression that a technological route is an inappropriate one.

Why do I say this? Precise scientific control of the brain is an illusion, on top of which there is the ever-present problem of side effects, the magnitude of which is of particular significance in the brain. For instance, as we have encountered previously, addiction appears to be intimately bound up with the mechanisms of action of some of the neuroenhancing drugs.[73] The speculative vistas served up in this area need to be tempered by empirical reality – what is and is not scientifically possible today and into the foreseeable future, as opposed to what might eventuate but for which there is no evidence of any description.

Specific ethical considerations implicit in attempts at neurally enhancing the brain include the balance between harms and benefits, curing and caring, respect for persons and human dignity, justice and concern for the poor, fairness and neighbour love.

73   Heinz et al., 'Cognitive Neuroenhancement'.

But there is a deeper Christian concern, and this is that dependence upon external forces to improve our moral responses is superficial and temporary. It leaves us as people untouched. Why act in way M rather than way N? Is it because we are convinced that M is preferable to N for substantial moral reasons, or because drugs acting on our brain move us in that direction? Dependence upon the support of drugs introduces a reductionist element into what we are as people. While there is a place for this therapeutically, it should be no more than an integral part of a broader therapeutic regime. For Christians the essence of the moral life is to love expecting nothing in return. Christians are to grow in maturity and understanding so that they do not rejoice in wrongdoing; they are not resentful, they do not insist on always getting their own way; they learn not to be envious, boastful, arrogant or rude. They are to be kind and to rejoice in all that is good and honourable and uplifting (1 Cor. 13). Living along these lines requires understanding of spiritual truths, of one's relationship to Christ, and of the reasons for seeking such a way of life. This is the essence of moral enhancement, and while one cannot ignore any neural basis for these attitudes, neither can modification of neural substrates produce them. The latter approach is far too restricted biologically, and also fails to account for the self-centeredness of human aspirations apart from divine intervention.

Humility and hubris

This leads to an emphasis on the centrality of humility not just as an individual virtue but as a central component in the way in which science and technology are to be approached.

Any Christian conception of humility will have as core features the importance of serving others and serving God, rather than serving ourselves. This will manifest itself in lowly acts of service, since we will not think of ourselves more highly than we ought (Rom. 12:3). The context for acting in these ways is that we are to be realistic about ourselves, knowing that we may on occasion be wrong. While these features illustrate how Christians themselves are to behave, and while they cannot automatically be imposed on others, they provide a broad base for good practice in the scientific realm.

If the general nature of these practices is accepted, we begin to appreciate how important humility is for scientific practice. Overconfidence in the reliability of scientific procedures and especially in our interpretation of them may be closely allied with overweening ambition, since the latter sometimes plays far too great a part in the way in which many scientists function. This in turn leads on some occasions to outright fraud and more commonly to over-interpretation of results. This is where hubris enters the picture, and with it excessive confidence in the speculations of scientists, especially where these go well beyond the results obtained.[74] We have encountered examples of this in the hyper-speculative digressions on cyborgs and posthumans and transhumanism in general, the assurance with which moral bioenhancement is put forward as a solution to a range of human problems, and in a far more restrained manner on the ARTs and regenerative medicine.

Faced with the variable speculative scenarios encountered here the realism inherent within Christian thinking entices us to question our motives, our grand theorizing and our incipient pride and arrogance. We may be wrong. We may be going in an unhelpful direction. We may be caught up in our own excessive ambition to reformulate human nature in our own image, or simply end up with ideas that please our own egos. And then in science as in every other area of life we all make errors of judgement, we all make mistakes, and sometimes our vistas turn out to be incomplete and unhelpful. Cautions such as these are directives not to be ignored, but sadly are frequently never taken into account. Honesty and objectivity are basic requirements in these exciting and enticing areas.

---

74   Zoë Corbyn, 'Misconduct Is the Main Cause of Life-Sciences Retractions', *Nature* 490 (2012), 21; D. Cyranoski, 'Retraction Record Rocks Community', *Nature* 489 (2012), 346–7; Jones and Whitaker, 'Scientific Fraud'.

CHAPTER 9

Pitfalls and hope in a technological world

Science, hope and values

Johannes Kepler (1571–1630) is said to have claimed that in his astronomical research he was merely 'thinking God's thoughts after Him'. It is far from clear whether Kepler did in fact make this claim, in spite of its frequent attributions to him. Notwithstanding this uncertainty, the statement has long been an important one for Christians in their work as scientists. They are indulging in an activity that is worthy of their calling as Christians and that enhances our appreciation of the extraordinary world created by God and which we experience everyday as human beings.

Kepler is said to have wanted to become a theologian, but for various reasons did not do so. In a letter written in 1595 to Michael Meastlin: 'I wanted to become a theologian; for a long time I was unhappy. Now, behold, God is praised by my work even in astronomy'.[1] While not in any way moving away from theology, Kepler recognized that as a scientist he had a great deal he could contribute to knowledge of God and his ways. He recognized the legitimacy of his science in providing accurate knowledge of the world, knowledge that had to be taken account of if one was to gain a working understanding of the universe and glimpses into the mind of the creator.

This resonates with me as a scientist deeply interested in and motivated by Christian imperatives. Throughout this book I have sought to base my thinking firmly in both reliable up-to-date science and Christian directives.

---

1   Johannes Kepler, 'Letter to Michael Maestlin (3 Oct 1595)', in *Johannes Kepler Gesammelte Werke* (1937), vol. 13, letter 23, l. 256–7.

To some this is a faulty base, since I am giving too much space to science and allowing it to dominate my Christian thinking. To others science alone should suffice; why bother with any Christian input? Especially within bioethics, secular utilitarian thinking is the framework of choice for many of the leading academics. This is not where I am coming from. For me, science and a Christian framework are essential, as long as I aim to follow Kepler on both fronts and 'throw away the nonsense and keep the hard kernel'.[2] Of course, that immediately raises the question of what constitutes the nonsense and the hard kernel. Both are matters of judgement, requiring knowledge and discernment, and a willingness to be prepared to accept one's own fallibility and on occasion gullibility.

But these are not exclusively cognitive considerations or merely academic matters for debate. They touch on deeply ingrained hopes and fears in all of us, since they impinge on what we are as human beings. How do we respond to one possibility after another: changing the characteristics and nature of our children or grandchildren; modifying the way in which we perceive reality; extending our lifespan way beyond present limits; providing us with an ever-expanding range of artificial parts; contemplating what perfection might look like in a baby or adult – the list can be added to almost indefinitely. The expectations appear to be endless, at least for those living with the financial means and resources to indulge in them, but even in these cases will they lead to happiness, contentment or fulfillment? Will they enlarge what we are as human beings, or will they simply change us in certain ways? The nature of the values available to guide these developments is essential.

The ethical issues encountered in these areas have intensely personal overtones, and have major repercussions for our hopes and fears. Each one of the developments alluded to in the previous paragraph has enticing as well as forbidding aspects, those that can be readily welcomed and those that appear to undermine cherished values. The dividing line will vary depending upon one's basic values and worldview, and yet there will be a

2   Kepler, 'Letter to Michael Maestlin (3 Oct 1595)', vol. 13, letter 68, l. 177.

dividing line and it is in this discernment that the balance will tilt towards fear or hope; hope challenged or hope regained.

In mapping a way forward I shall discuss three overriding considerations that might be of value in gaining a sense of perspective. These are the relationship between God's care and human care, how one might go about discerning landmarks for what it means to be human, and how to best approach an uncertain and unknown future.

God's care versus human care

It is common to encounter the apparent contrast between human and divine control, with its potential to generate conflict. The reason for this is that, as human control has become increasingly precise, the questions it raises have also become increasingly precise. This precision forces us to consider whether it is expected that God's control will be expressed in comparably precise terms, and whether God micro-manages the lives of human beings.

Biblical justification for the latter appears to be found in the words of Jesus when he made the comment that 'even the hairs of your head are all counted' (Luke 12:7). Is this to be taken literally, and is the modern equivalent the counting of every skin cell, mucosal cell, nerve cell or gene? The context within which Jesus made this comment was that of reassuring his followers that God would protect them from those set on destroying them and their faith, thereby separating them from Christ and the new way of life enshrined in him. To make his point he drew a contrast between the comparative value of human beings and sparrows. Since God's concern extends to sparrows, surely he contended it would also encompass his followers since their value is far greater (Luke 12:4–7). From this it followed that they were not to be afraid when confronted by those intent on destroying them. God knew them not just as nebulous individuals, but also down to the level of the hairs on their heads. Later in the same dialogue, Jesus told them not to worry about what they would eat or wear. Since God's care extends to birds like ravens and flowers like lilies, in spite of their brief and fragile existences, his care for his disciples should be self-evident. To emphasize this, he reminded them of the obvious fact that worrying would not enable them to add to their height or extend their lives (Luke 12:22–31).

This discourse provides a view of God's care. Except in some illnesses it is irrelevant how many hairs people have, and even then it is not generally the most important issue in life. What would be the value of God controlling every single hair on a person's head (or body)? It is far more important to know that he cares, as evidenced by his concern for even the detailed aspects of people's lives, including their genes, cells, organs and body systems. The exact number of hairs people have or even the precise formulation of their genes (important though the latter may be) are not of supreme importance, but rather the manner in which they are able to cope with the hairs and genes they do or do not have. Coping along these lines is to occur against a background of God's care and concern for them as people.

It is for this reason that this section has as its title: 'God's care versus human care'. As the range of human abilities is extended by scientific achievements, the pressing issue is the way in which humans utilize these abilities. Human control over biological and other processes is increasing, but it will never be complete. Nevertheless, regardless of how extensive or otherwise it is, the challenge is to direct these abilities in ways that will help individuals and communities, in particular towards those in greatest need. If debate centered more on care than on control, it would demonstrate the centrality of a God-oriented perspective rather than push this perspective to the periphery.

God's care is a reminder that what is important is the way in which we care for others; not the way in which we attempt to control others and lord it over them. This, in turn, will determine the manner in which technology is used, to benefit others, rather than to one's own aggrandizement. The character of the lives of humans, both as individuals and as communities, demonstrates the extent to which they image God and do or do not honour him. This is demonstrated by attempts to add value to the life experiences of others, especially the needy and downtrodden, over against using one's abilities and facilities simply to enhance one's own comforts. Considerations along these lines are integral to the responsibility of human beings, as those made in the image of God.

A temptation is to see new technological developments as threats to the standing of human beings, and particularly to the manner in which God operates. In succumbing to this way of thought, human beings downplay their responsibility for directing these developments and advocating for them. Christians may be uncomfortable about certain trends in society, including the direction of technological implementation, and that is to be expected in any pluralist, largely secular society. However, their task is to use the opportunities presented to them to advocate for a 'better way'. But what might this 'better way' entail?

Who determines the future?

A core consideration is provided by the description of the stature of human beings that appears in both the Old and New Testaments:

> What are human beings that you are mindful of them, mortals that you care for them? Yet you have made them a little lower than God, and crowned them with glory and honour. You have given them dominion over the works of your hands; you have put all things under their feet. (Ps. 8:4–6)

> What are human beings that you are mindful of them, or mortals that you care for them? You have made them for a little while lower than the angels; you have crowned them with glory and honour, subjecting all things under their feet. Now in subjecting all things to them, God left nothing outside their control. As it is, we do not yet see everything in subjection to them, but we do see Jesus [...] (Heb. 2:6–9)

It is interesting reading these words against a backdrop of the frequently expressed criticism that scientists overreach themselves and act as if they were gods, masters of their fate and of the destiny of all around them. These days it is biomedical scientists who may come in for this criticism. And yet the biblical writers who penned these words, admittedly writing

aeons before the advent of the modern technological era, did not appear
to think in anything resembling these terms. For them, the grandeur of
the human condition is a given, even while acknowledging that much has
gone wrong, that human beings have lost their way, and that they need
direction. While these words taken in isolation do not pretend to provide
a complete theology of the human situation, they are reminders of crucial
juxtapositions: our elevated stature alongside our mortality; our authority
over the creation alongside our own need to be subject to a higher authority; the way in which we are cared for by God, over against the care and
control we are to exercise over the creation and over others in the human
community. These words point towards humble responsibility towards,
rather than arrogance or exalted dominion over, the natural world.

It is in the working out of these juxtapositions that humans live truly
human existences, powerful and yet humble, leaders and yet willing to be
led, making use of potentially transformative technologies but accepting
their limitations and their own proclivity to misuse them and be led astray
by them. Humans should not shun the possibilities inherent in being designers of the future, but immense wisdom, discernment and understanding
are crucial.[3]

It is necessary to take science seriously, and to appreciate its role within
God's economy. No matter how much science can be misused, whether in
the realm of overweening technology or in the manner in which it can be
transformed into philosophical scientism, its importance is unquestioned. It
provides means of overcoming one limitation after another – illness, disease,
poverty. It provides power to make life more comfortable, more fulfilling
and more purposeful. To constantly stress its misappropriation and the
ways in which it can be employed to mislead is unhelpful. The challenge
from a Christian viewpoint is to have sufficient faith to recognize God's
hand in the blessings that scientific achievements bring, and to discern his
directives when faced with the distractions and dangers that sometimes
follow in its wake. This is a start, but what of the future? This might be far
more forbidding than anything we have experienced up to now.

3   Jones, *Designers of the Future.*

Does it make sense to think human beings can 'choose the future'? Or will the future arrive unheralded, wanted or unwanted, on our doorsteps? Is it feasible to prevent one kind of future while encouraging another kind, and does this lie solely in the hands of scientists and clinicians? Repeatedly scientific developments emerge into social consciousness with little previous debate over the likely repercussions for society. No wonder there is so much angst among groups like theologians and social scientists, as they picture the human race going down paths mapped out by scientists and, so it seems, entirely at their behest.

Of the many examples, consider the following concerning Sir Robert Edwards, and his pioneering work in bringing IVF to fruition and the clinic. Some of the social and theological issues arising from IVF and the reproductive revolution were discussed in chapter 3, and all have taken place following the introduction of IVF and the ARTs in general. What is relevant here is that it was Edwards, the physiologist, and Patrick Steptoe, his obstetric colleague, who drove these developments.

In the eyes of many, Edwards was attempting to do something that was impossible scientifically and untenable ethically. He was motivated by one simple dictum: 'the most important thing in life is having a child'.[4] Edwards combined the fascination of the basic scientist to understand more about human fertilization, the drive of the applied scientist to help those with infertility issues, and a serious commitment to ethical debate. In each area he challenged accepted thinking and attitudes.

In his 1989 book, *Life before Birth*, he reflected on the debate raging over the work he was doing, and the opposition he had encountered, including libel actions he took out against, among others, the British Medical Association.[5] His thinking even then extended well beyond IVF as he tackled issues to do with the moment of fertilization, embryo donation,

---

4   BBC News, 'Test-Tube Baby Pioneer Sir Robert Edwards Dies', (10 April 2013) <http://www.bbc.co.uk/news/uk-england-cambridgeshire-22091873> accessed 26 June 2013.
5   Robert Edwards, *Life before Birth* (London: Hutchinson, 1989).

embryo freezing, the prenatal diagnosis of genetic defects, sex selection, stem cells and research on embryos.

The extent of his involvement in ethical debate was unusual for a practising scientist, and even more so as he engaged with politicians, philosophers and theologians. He wanted society to take informed decisions, but was constantly disappointed: 'all we've produced so far is acrimony, arguments, amendments, confusion'.[6]

Edwards had a penchant for headline-grabbing quotes. In 1999 he told an International Fertility Conference in France, 'Soon it will be a sin for parents to have a child that carries the heavy burden of genetic disease. We are entering a world where we have to consider the quality of our children'.[7] In 2003 he told a reporter, 'I wanted to find out exactly who was in charge, whether it was God himself or whether it was scientists in the laboratory'.[8] And what he discovered was that 'It was us'. In making this statement he was not speaking with the voice of a scientist, in addition to which its bravado was unhelpful.

There is no doubt that Edwards changed the world. The 5 million individuals alive today as a result of IVF would not have existed, and the number is growing every day. The social effects of IVF are legion: older women flock to use it; the donation of eggs and sperm, let alone embryos, in conjunction with IVF has become commonplace; PGD for the detection of genetic defects in embryos has become possible. The benefits and drawbacks of each of these procedures continue to be ardently debated as questions around the moral status of the embryo have moved out of academic debating chambers into everyday clinical practice.

It is no wonder that the announcement of the belated award of the Nobel Prize to then eighty-five-year-old Edwards in 2010 was greeted with a mixture of delight and scorn. At the time a Vatican spokesman commented that without Edwards there would be no market for human eggs, and no

6   Edwards, *Life before Birth*, 180.
7   Lois Rogers, 'Having Disabled Babies Will Be "Sin," Says Scientist', *Sunday Times* (4 July 1999).
8   'Interview with Robert Edwards', *The Times* (24 July 2003).

freezers full of embryos,[9] while another declared that IVF 'ignores all the problems of ethics and stresses that man can be reduced from a subject to an object'.[10] Over against this the Nobel Committee declared that his work had brought joy to infertile people all over the world.[11]

Both stances should be given due weight. Edwards cannot be accused of ignoring ethics; he was discussing reproductive ethics long before most other voices, including theological ones, joined the fray. His permissive stance was not to everyone's liking, and there can be no doubt that the ability to isolate and manipulate human embryos in the laboratory has opened numerous doors both scientifically and socially (chapter 3). Edwards had truly opened Pandora's box, although for many infertile couples that has been more positive than negative.

Edwards forced a fundamental rethinking of how we define fertilization, how we think about very early embryos in the laboratory, and how we balance the needs of the infertile against our responsibilities towards these very earliest versions of ourselves. In bringing together his innovative science and his fearless ethical thinking he set the scene for one of the most revolutionary procedures in biomedicine to emerge in the latter part of the twentieth century.

In the 1960s, IVF lay well and truly in the future. For most it did not exist. There were no colloquia devoted to discussing it, and only a few lone theologians had it in their sights, and for them it represented exceedingly forbidding territory.[12] Arguments included the immoral nature of producing the first child conceived in a laboratory since one could not predict in advance how the experiment would turn out. While it would be unfair to

9   BBC News, 'Vatican Official Criticises Nobel Win for IVF Pioneer'.
10 Catholic News Agency (CNA), 'Vatican Health Experts "Dismayed" by Nobel Prize for IVF Co-Developer', (5 October 2010) <http://www.catholicnewsagency.com/news/vatican-health-experts-dismayed-by-nobel-prize-for-ivf-co-developer/> accessed 26 June 2013.
11 'The 2010 Nobel Prize in Physiology or Medicine – Press Release', *Nobelprize.org* (4 October 2010) <http://www.nobelprize.org/nobel_prizes/medicine/laureates/2010/press.html> accessed 26 June 2013.
12 Ramsey, 'Shall We "Reproduce"? I.'; Ramsey, 'Shall We "Reproduce"? II.'.

deride such a stance, scientists simply undertook the work, driven as they were by the desire to help the infertile, and also by the inherent fascination of accomplishing something in reproductive biology that had never been accomplished before. The scientific challenge was, and always is, formidable.

The example provided by Edwards is instructive. As outlined in chapter 3, the ongoing effects of the reproductive revolution are immense. The implications for society have been diverse and, depending upon one's perspective, range from ambivalent to disconcerting, from challenging to highly problematic. They have opened the doors to new family structures as gamete donation within and between generations has become feasible, while the availability of embryos in laboratory settings has made possible an increasing range of manipulatory endeavours including the production of human-animal hybrid embryos, and forms of cloning. PGD, with its reliance upon the availability of IVF, has in its turn made possible a new genre of genetic analysis with all the ethical and theological implications associated with the elimination of genetic conditions at the embryonic stage.[13]

The dominant take home message from this one example is that scientific advance will continue to lie within the province of scientists, although ideally the applications of the resulting procedures and their availability or non-availability within a society will be the subject of wide multidisciplinary debate. It is highly unlikely that it will prove feasible, let alone desirable, for theologians and others to have input into the scientific process itself. It is here that major concerns emerge on account of the reductionist methodology of science and the likelihood that this will reign supreme in ethical and policy decision-making. The concern is that the embryo (in this instance) could become little more than an experimental tool. Cautious voices emanating from theologians and others find difficulty in getting a hearing, since they give the appearance of being outsiders looking in at the scientists, and speaking a foreign language. The impression is given that they

13   J. Malek, 'Deciding against Disability: Does the Use of Reproductive Genetic Technologies Express Disvalue for People with Disabilities?', *Journal of Medical Ethics* 36/4 (2010), 217–21; J. A. Robertson, 'Extending Preimplantation Genetic Diagnosis: The Ethical Debate. Ethical Issues in New Uses of Preimplantation Genetic Diagnosis', *Human Reproduction* 18/3 (2003), 465–71.

are interlopers expressing a viewpoint that comes from a different discipline that is seeking to impose its perspective on the prevailing science. Doubts along these lines do not apply only to theological input, although concerns may be greater in this case than with social scientists or gender theorists due to the religious overtones. However, in each society, a way has to be found of moderating the ambitions of scientists with the more broadly based interests of other disciplines, on the understanding that the dynamics in different societies will lead to a range of solutions governing what can and cannot be done to embryos.[14] This provides a salutary reminder to naysayers that everything that is claimed to be possible scientifically will not eventuate, either politically or scientifically.

Predicting the future?

To look into the future is a hazardous, even unwise, pastime. The chances of being appallingly wrong are very high. At the level of detailed analysis it is not possible to know with any accuracy what questions will arise before they have arisen at a scientific level. The direction that scientists take is so often determined by the in-built rationale of the science itself, and not by the pronouncements of ethicists, theologians or governmental agencies. Further, the ability to predict precisely what will turn out to be of scientific value is poor.

Imagine a world many years in the future, 2080 in fact. There is nothing special about this particular year any more than 1984 or 2001 were special. In fact, as we look back on those other two years, we probably think of them as mundane. And yet, when the novel, *1984*[15] and the film *2001: A Space Odyssey* were written, each in their very different ways looked forward to a future time when life had taken on totally different dimensions,

14 C. R. Towns and D. G. Jones, 'Stem Cells: Public Policy and Ethics', in M. Ruse and C. A. Pynes, eds, *The Stem Cell Controversy: Debating the Issues* (New York, NY: Prometheus Books, 2004), 329–41; C. R. Towns and D. G. Jones, 'Stem Cells: Public Policy and Ethics', *NZ Bioethics Journal* 5 (2004), 22–8.

15 George Orwell, *Nineteen Eighty-Four* (London: Secker and Warburg 1949).

unknown, unparalleled and grimly strange. These were indeed 'brave new worlds'. The year 2080 will be no different and yet it is difficult to believe that it will not be a world of biotechnological control and biomedical manipulation. Imagine what 2080 might look like:

> There will be few books since everything will be available online. Information will be nearly instant, although the problems with this will be considerable. A new industry will have started up to assist in sorting out useful from useless information. The information overload will be immense. No one will wear spectacles since all forms of eye defects will be treated via laser surgery. In the same way there will be no surgery in the old fashioned sense, since laser and remote controlled surgery will have taken over. Even hospitals will almost have ceased to exist; day surgeries will serve almost every need in the community. Alzheimer's disease will be a phenomenon of the past, with people living to 120 years as a rule and without the threat of dementia hanging over their heads. Infertility will have been conquered and few people will have babies naturally. Why should they, when what we know as IVF and PGD will be so routine, cheap and easy to administer? Clones will exist and will be treated as perfectly ordinary members of society. Most are anonymous, just as those conceived by IVF are anonymous today. Over 30 per cent of the population will have artificial parts of some description – for example, brain implants, with large tracts of the frontal lobes of their brains having been replaced by artificial devices, to make them more loving and considerate.
>
> Of course, there are problems. It is difficult to know what is real, and the idea of the natural seems to have disappeared almost entirely. Most people at some stage of their lives have organs like the heart and kidneys replaced by small artificial devices that fit neatly inside their bodies. They don't have to wait for these organs to malfunction to have them replaced; the implants are said to function better than the natural ones. Legs and joints are routinely replaced by prostheses, and few people over the age of 45 have any of their own natural joints. There are no individuals with Down Syndrome or any other genetic condition, since only healthy babies are born. However, people still die, even if many are centenarians when they do.

Two points are significant. The first is that the origins of all these possibilities are already with us. Think of prostheses, laser surgery in its various forms, the reproductive technologies, gradually emerging therapies to slow the onset of dementia in Alzheimer's disease and the increasing sophistication of genetic analysis. The second is that human beings, no matter how changed in some respects, will be substantially similar to people today. There is no evidence to suggest that basic human aspirations have changed dramatically over the past few hundred years or that they differ greatly between widely disparate cultures. It is true that expectations change as childhood mortality changes and as health status changes. But much that is fundamental to human life remains constant and will likely do so into the future.

Between our world and that of 2080 is a continuum. There will be no sudden point when our world becomes that far-off world. There is continuity between the two. To some extent the world of 2080 is already with us. If the world of 2080 is described as a nightmare, there are hints of that nightmare even now. It will not depend upon a wayward group of scientists getting out of control. Ordinary people going about their day-to-day business are already partaking in this transition.

The numerous biological and medical developments I have been dealing with throughout these chapters are everyday developments. They are part of the warp and woof of daily existence. There is one small change here and another small change there, most of the changes being welcome ones – at least for some sections of the population. We are not surrounded by Frankenstein-like scientists in white coats working to an agenda aimed at conquering the world. They are ordinary people doing ordinary jobs. Their normality is their most striking feature.

It is therefore the ordinary and frequently mundane decisions that call for our attention. Each one is worth attending to diligently and carefully, since each is significant for those called to be stewards of God's creation in the laboratory, office, consulting room or committee room. It is unfortunate that some Christians see Armageddon-like vistas, and hence find these small but cumulatively massive changes difficult to come to terms with. The result is that they cling to the image of the malevolence of science and scientists, at least in selected areas. Unfortunately, all too often this approach leads to a tendency to exaggerate issues in order to provide grand and forbidding vistas to combat, whether cloning, embryonic stem cells, eugenics or designer babies. Such an approach lends itself to paralysis in the face of the unknown.

The future scenario given here is not to be taken literally, since most of the developments will not emerge as predicted. Neither is it to suggest blanket agreement with any of these developments. My openness to the

future is not one of unalloyed welcome and acceptance. Its thrust is to prepare us for future movements, the trajectory of which can be glimpsed from
the general directions of biomedical scientific investigations. I have made
no attempt to take note of other trends that may have major consequences
for medical technology, from climate change to political upheaval, and of
course I have no idea what may arise out of the blue. I consider there are
no grounds to reject these general directions from a Christian perspective; what is crucial is the assessment of each one in light of the effects its
application might have on people and communities.

Forging hope rather than fear

In the final chapter of the edited volume: *A Glass Darkly: Medicine and
Theology in Further Dialogue*, I wrote the following:

> This book has picked up on these limitations and uncertainties in our knowledge and
> understanding when considering frighteningly complex and pressing contemporary
> matters in the biomedical realm. Our contention has been that it is incumbent upon
> us to steer a course between almost complete assurance as to the rightness of our
> cause and despair that helpful guidance will ever be found. We are only too aware that
> we do not see as clearly and unambiguously as we would like, and that all too often
> we are dealing with puzzling reflections rather than seeing with 20/20 vision. But
> we have also proposed ways forward for all of us – scientists, clinicians, theologians
> and those seeking to live out their lives in this contested territory. We have sought
> to map a course that takes seriously the diverse factors impinging upon bioethical
> decision-making, a course built upon a realistic appreciation of what Christian hope
> can mean for those of us living in a world dominated by biomedical technology.[16]

Earlier in that chapter I had alluded to some of the writings of Verhey,
one of his conclusions being that, 'The memory of Jesus does not provide
any neat and easy resolution to such conflict. It does not usher in a new

16   Jones, 'Conclusion', 238.

heaven and a new earth, either. Here and now there is ambiguity'.[17] This was not a conclusion of agnosticism or despair but a realistic assessment of how those functioning within a Christian framework are to approach biomedical issues. Go as far as possible, but do not expect to resolve with certainty the many conflicts inherent within the issues. This is not a sign of failure but of the nature of the issues.

Genuine dialogue is vital, both within the Christian community, and between members of that community and those outside it. In both instances, there is diversity, and diversity can readily generate conflict. Diversity can also be a strength, if carried out in a spirit of humility with all prepared to accept at least the validity of a range of expertise and insights. This can be difficult when insights give the impression of moving in opposite directions.

There has to be honesty in interpreting scientific investigations, ensuring that they are not over-interpreted. By the same token there is to be no denigration of what appear to be legitimate interpretations. Alongside these considerations there is to be honesty in the manner in which theological and biblical data are interpreted – once again these are not to be over-interpreted. There is no easy road to success and efforts will frequently fall short of our expectations and hopes. This will often entail walking in the midst of diversity and uncertainty. Nevertheless, science can be used productively even in such unnerving territory, and Christians should use it in this way.

There is also a down-to-earth realism in Christianity that should serve to ground Christians as they walk these tenuous paths. Consider the following contrast. In chapter 7, some of the claims and hopes of transhumanism were touched upon. In line with what emerged there, Aubrey de Grey, the well-known Cambridge transhumanist, believes we can stop the physical deterioration that normally accompanies ageing, thereby giving to people the prospect of living to be 1,000 years of age.[18] This he says is founded

---

17   Verhey, 'What Makes Christian Bioethics Christian?', 313.
18   Aubrey D. N. J. de Grey et al., 'Time to Talk SENS: Critiquing the Immutability of Human Aging', *Annals of the New York Academy of Sciences* 959/1 (2002), 452–62.

not on faith but on science.[19] For de Grey, who is the co-founder of the SENS Research Foundation,[20] getting old is the biggest health crisis facing the world. Why? Because, according to him, two-thirds of all deaths are caused by ageing. Hence, the defeat of ageing is the most compelling challenge facing humankind, with its prospect of saving 100,000 lives a day.

Transhumanists of the de Grey ilk are committed to curing death and ushering in a form of physical immortality, in which illness will have been banished to the past and all people who are alive will be healthy. The prospects opened up by these scenarios are staggering – apparently overpopulation can be avoided, and one imagines happiness will be undiluted and unending. The worldview enshrined by this allegedly scientific-based view has nothing in common with any Christian worldview, and takes scientific achievements to the nth degree. Extreme as this view is, and unlikely as it is to eventuate, the reactions it prompts in many are ones of fear and foreboding.

In spite of the concerns expressed by exponents of these views over the unnecessary deaths each year, they do nothing to face up to the inequalities of the present world. For instance, the life expectancy gap between rich and poor people in England and Wales is widening, despite years of government action. Extensive efforts have failed to reduce the wide differential, which can be ten years or more depending on socioeconomic background.[21] These figures are mirrored in other comparable societies. It has been estimated that if African-American people in the US had the same mortality rates as Caucasian people there would have been 800,000 fewer deaths over a

---

19   Caspar Llewellyn Smith, 'Aubrey De Grey: We Don't Have to Get Sick as We Get Older', *The Observer* (1 August 2010).

20   See <http://www.sens.org>.

21   J. Meikle, 'Life Expectancy Rises in UK but North-South Divide Widens', *The Guardian* (2011) <http://www.guardian.co.uk/society/2011/jun/08/life-expectancy-north-south-divide-widens> accessed 26 June 2013; Statistical Directorate, 'Statistical Bulletin, Life Expectancy 2007–2009', Welsh Assembly Government (24 November 2010) <http://wales.gov.uk/docs/statistics/2010/101124sb942010en.pdf> accessed 26 June 2013.

decade.[22] These health inequalities clearly highlight unfairness that in turn costs lives, damages people's health and stunts educational opportunities. The wider causes of ill health are poverty and poor housing.

The ethical and theological lessons that arise from these figures should be obvious. They are a blot on our world, with far greater inequalities than these if we compare different societies. For instance, the differential life expectancy between women in Sierra Leone on the one hand and Japan on the other is forty years.[23] While the causes are legion, the means of rectifying these differentials, at least to a reasonable extent, are available. Scientific knowledge is part of the answer, since awareness of the importance of clean water, adequate sewerage, vaccination and antibiotics has been present for many years, with HIV/AIDS infection adding to the burden of those with the lowest life expectancies in recent years. Health inequality should be an urgent priority, since it can be remedied, or at the very least ameliorated. It touches the lives of real people who are loved by God, and who should be loved by others. For Christians it brings together their commitment as Christians and their ability to utilize scientific abilities in the service of others.

These two items bring us face to face with the powers and prospects of scientific investigations. If any of the developments touched on in either item are to be realized, even partially, they will require considerable skill. However, the efforts put into these developments highlight the priorities we bring to them as human beings. From a Christian stance, efforts should be directed into alleviating the multidimensional plight of those in dire need now. These are the weak and the poor of the contemporary world, not the rich who wish to live a few, or even many, years longer. Christians look forward to a world in which illness and inequality are being addressed, using all the means available. Christians are to work towards rectifying the groaning creation, with its overweening disease and dis-ease. To accomplish

---

22   Robert A. Hummer and Juanita J. Chinn, 'Race/Ethnicity and US Adult Mortality', *Du Bois Review: Social Science Research on Race* 8/1 (2011), 5–24.

23   World Health Organization, 'Life Expectancy by Country', <http://apps.who.int/ gho/data/node.main.688?lang=en> accessed 26 June 2013.

this, even in small measure, requires an increase in understanding as well as a determination to rectify what can be rectified. Science is pivotal in such stirrings. While it is not the one and only answer, it is a significant contributory factor in bringing about these possibilities.

What is required is a balance between scientific contributions and ethical drivers such as justice, honesty, integrity and equity. Working for justice, and using scientific means as appropriate towards these ends, stems from Christian aspirations. A theological framework centers on hope and trust in God's good purposes. It recognizes the limited character of human abilities and the reality of human fallenness and rebellion against God, while being able to look beyond human abilities in hope, even as the crucial contributions humans have to make are recognized.

The interface between theology and science is not simply a theoretical and academic one, it is the commitment of redeemed people who are made new in Christ, to do God's will and usher in hints and touches of God's kingdom. For those in the sciences and with active involvement in science, an important element will be the incorporation of scientific understanding into their deliberations, as they attempt to use it and its results for the good of humankind and hence for the glory of God.

Bibliography

'The 2010 Nobel Prize in Physiology or Medicine – Press Release', *Nobelprize.org* (4 October 2010) <http://www.nobelprize.org/nobel_prizes/medicine/laureates/2010/press.html> accessed 26 June 2013.

Advisory Committee on Assisted Reproductive Technology, 'Use of Gametes and Embryos in Human Reproductive Research: Summary of Submissions', Ministry of Health (September 2007) <http://www.acart.health.govt.nz/moh.nsf/pagescm/6730> accessed 17 June 2013.

Agar, Nicholas, 'Whereto Transhumanism? The Literature Reaches a Critical Mass', *Hastings Center Report* 37/3 (2007), 12–17.

——, 'Why Is It Possible to Enhance Moral Status and Why Doing So Is Wrong?', *Journal of Medical Ethics* 39/2 (2013), 67–74.

Andersen, R, 'Why Cognitive Enhancement Is in Your Future (and Your Past)', *The Atlantic* (6 February 2012) <http://www.theatlantic.com/technology/archive/2012/02/why-cognitive-enhancement-is-in-your-future-and-your-past/252566/> accessed 23 June 2013.

Anderson, Hamish, 'Preimplantation Genetic Diagnosis: From Clinic to Eugenic Fears and Disability Concerns', (Thesis, Bachelor of Medical Science with Honours, University of Otago, 2012) <http://hdl.handle.net/10523/2335> accessed 17 June 2013.

Arlidge, John, 'Scientists "Able to Create Human Clone"', *The Guardian* (26 February 1997), 6.

Azari, Nina P, 'Neuroimaging Studies of Religious Experience: A Critical Review', in Patrick McNamara, ed., *Where God and Science Meet: How Brain and Evolutionary Studies Alter Our Understanding of Religion* (Westport, CT: Praeger Publishers, 2006), 33–54.

Balen, Adam H., and Anthony J. Rutherford, 'Managing Anovulatory Infertility and Polycystic Ovary Syndrome', *BMJ* 335/7621 (2007), 608–11.

Banwell, B. O., *The Illustrated Bible Dictionary* (Leicester: Inter-Varsity Press, 1980).

Barns, Ian, 'Debating the Theological Implications of New Technologies', *Theology and Science* 3/2 (2005), 179–96.

Bartke, A., J. C. Wright, J. A. Mattison, D. K. Ingram, R. A. Miller, and G. S. Roth, 'Longevity: Extending the Lifespan of Long-Lived Mice', *Nature* 414/6862 (2001), 412.

Baskin, J. H., J. G. Edersheim, and B. H. Price, 'Is a Picture Worth a Thousand Words? Neuroimaging in the Courtroom', *American Journal of Law & Medicine* 33/2–3 (2007), 239–69.

Batsch, N. L., M. S. Mittelman, and Alzheimer's Disease International, *World Alzheimer Report 2012* (London: Alzheimer's Disease International, 2012) <http://www.alz.co.uk/research/world-report-2012> accessed 23 June 2013.

BBC News, 'Test-Tube Baby Pioneer Sir Robert Edwards Dies', (10 April 2013) <http://www.bbc.co.uk/news/uk-england-cambridgeshire-22091873> accessed 26 June 2013.

——, 'Vatican Official Criticises Nobel Win for IVF Pioneer', (4 October 2010) <http://www.bbc.co.uk/news/health-11472753> accessed 17 June 2013.

Bellamy, J. Stephen, 'Evangelicals and Embryology: Responses to the Human Fertilisation and Embryology Bill', in D. Gareth Jones and R. John Elford, eds, *A Glass Darkly: Medicine and Theology in Further Dialogue* (Bern: Peter Lang, 2010), 157–90.

Berry, Robert James, *God and the Biologist* (Leicester: Apollos, 1996).

Beyhan, Zeki, Amy E. Iager, and Jose B. Cibelli, 'Interspecies Nuclear Transfer: Implications for Embryonic Stem Cell Biology', *Cell Stem Cell* 1/5 (2007), 502–12.

Birbaumer, N., 'Breaking the Silence: Brain-Computer Interfaces (BCI) for Communication and Motor Control', *Psychophysiology* 43/6 (2006), 517–32.

Bird, A., 'Perceptions of Epigenetics', *Nature* 447/7143 (2007), 396–8.

Blair, R. J., 'The Amygdala and Ventromedial Pre-Frontal Cortex in Morality and Psychopathy', *Trends in Cognitive Science* 11 (2007), 387–92.

Blumenthal-Barby, J. S., 'Between Reason and Coercion: Ethically Permissible Influence in Health Care and Health Policy Contexts', *Kennedy Institute of Ethics Journal* 22/4 (2012), 345–66.

Board for Social Responsibility (Working Party on Human Fertilization and Embryology), *Personal Origins* (London: CIO Publishing, 1985).

Board for Social Responsibility of the Church of England, *Cybernauts Awake!* (London: Church House Publishing, 1999).

Bolte Taylor, Jill, *My Stroke of Insight: A Brain Scientist's Personal Journey* (London: Hodder, 2006).

Boomsma, Dorret, Andreas Busjahn, and Leena Peltonen, 'Classical Twin Studies and Beyond', *Nature Reviews Genetics* 3/11 (2002), 872–82.

Bostrom, Nick, 'A History of Transhumanist Thought', *Journal of Evolution and Technology* 14/1 (2005), 1–25.

——, 'Transhumanist Values', *Review of Contemporary Philosophy* 4/1–2 (2005), 87–101.

Brookmeyer, R., E. Johnson, K. Ziegler-Graham, and H. M. Arrighi, 'Forecasting the Global Burden of Alzheimer's Disease', *Alzheimer's & Dementia* 3/3 (2007), 186–91.

Brown, Rachel, and Joyce Harper, 'The Clinical Benefit and Safety of Current and Future Assisted Reproductive Technology', *Reproductive BioMedicine Online* 25/2 (2012), 108–17.

Brunet, A., S. P. Orr, J. Tremblay, K. Robertson, K. Nader, and R. K. Pitman, 'Effect of Post-Retrieval Propranolol on Psychophysiologic Responding During Subsequent Script-Driven Traumatic Imagery in Post-Traumatic Stress Disorder', *Journal of Psychiatric Research* 42/6 (2007), 503–6.

Brunner, Han G., M. Nelen, X. O. Breakefield, H. H. Ropers, and B. A. Van Oost, 'Abnormal Behavior Associated with a Point Mutation in the Structural Gene for Monoamine Oxidase A', *Science* 262 (1993), 578–80.

Bryant, John, and John Searle, *Life in Our Hands* (Nottingham: Inter-Varsity Press, 2004).

Buchanan, Allen, *Better Than Human: The Promise and Perils of Enhancing Ourselves* (Oxford: Oxford University Press, 2011).

——, 'Enhancement and the Ethics of Development', *Kennedy Institute of Ethics Journal* 18 (2008), 1–34.

Callaway, Ewen, 'Murderer with "Aggression Genes" Gets Sentence Cut', *New Scientist* (3 November 2009) <http://www.newscientist.com/article/dn18098-murderer-with-aggression-genes-gets-sentence-cut.html> accessed 7 April 2013.

——, 'Shocks to the Brain Improve Mathematical Abilities', (16 May 2013) <http://www.nature.com/news/shocks-to-the-brain-improve-mathematical-abilities-1.13012> accessed 23 June 2013.

Cameron, Nigel M. de S., *The New Medicine: The Revolution in Technology and Ethics* (London: Hodder & Stoughton, 1991).

Campbell, A. V., *The Body in Bioethics* (London: Routlege, 2009).

Campbell, A. V., G. Gillett, and D. G. Jones, *Medical Ethics* (Melbourne: Oxford University Press, 2005).

Cartlidge, Edwin, 'Meet the World's First Transhumanist Politician', *New Scientist* (18 September 2012) <http://www.newscientist.com/article/mg21528826.100-meet-the-worlds-first-transhumanist-politician.html > accessed 12 April 2013.

Caspi, A., J. McClay, T. E. Moffitt, J. Mill, J. Martin, I. W. Craig, A. Taylor, and R. Poulton, 'Role of Genotype in the Cycle of Violence in Maltreated Children', *Science* 297/5582 (2002), 851–4.

Catholic News Agency (CNA), 'Vatican Health Experts "Dismayed" by Nobel Prize for IVF Co-Developer', (5 October 2010) <http://www.catholicnewsagency.

com/news/vatican-health-experts-dismayed-by-nobel-prize-for-ivf-co-developer/> accessed 26 June 2013.

Caulfield, T., and A. L. McGuire, 'Direct-to-Consumer Genetic Testing: Perceptions, Problems, and Policy Responses', *Annual Review of Medicine* 63 (2012), 23–33.

Chorney, M. J., K. Chorney, N. Seese, M. J. Owen, J. Daniels, P. McGuffin, L. A. Thompson, D. K. Detterman, C. Benbow, and D. Lubinski, 'A Quantitative Trait Locus Associated with Cognitive Ability in Children', *Psychological Science* 9/3 (1998), 159–66.

Christian Institute, 'The Sanctity of Life', <http://christian.org.uk/briefingpapers/sanctityoflife.htm> accessed 29 May 2013.

Christmas, David, Colin Morrison, Muftah S. Eljamel, and Keith Matthews, 'Neurosurgery for Mental Disorder', *Advances in Psychiatric Treatment* 10/3 (2004), 189–99.

Churchland, Patricia S., *Neurophilosophy: Toward a Unified Science of the Mind-Brain* (Cambridge, MA: MIT Press, 1986).

Churchland, Paul M., *The Engine of Reason, the Seat of the Soul* (Cambridge, MA: MIT Press, 1995).

Clark, Andy, *Natural-Born Cyborgs: Minds, Technologies, and the Future of Human Intelligence* (Oxford: Oxford University Press, 2003).

Clarke, Kevin, 'O'Malley Condemns Cloning Breakthrough', *America: The National Catholic Review* (15 May 2013) <http://americamagazine.org/content/all-things/omalley-condemns-cloning-breakthrough> accessed 17 June 2013.

Clément, K., C. Vaisse, N. Lahlou, S. Cabrol, V. Pelloux, D. Cassuto, M. Gourmelen, C. Dina, J. Chambaz, and J.-M. Lacorte, 'A Mutation in the Human Leptin Receptor Gene Causes Obesity and Pituitary Dysfunction', *Nature* 392/6674 (1998), 398–401.

Cobbe, Neville, 'Cross-Species Chimeras: Exploring a Possible Christian Perspective', *Zygon* 42/3 (2007), 599–628.

Coker, Robina, *Complementary and Alternative Medicine: Should Christians Be Involved?* (London: Christian Medical Fellowship, 2008).

Cole, Robert, and D. Gareth Jones, 'Testing Times: Do New Prenatal Tests Signal the End of Down Syndrome?', *New Zealand Medical Journal* 126/1370 (2013), 96–102.

Cole-Turner, R., *The New Genesis: Theology and the Genetic Revolution* (Louisville, KY: Westminster/John Knox Press, 1993).

——, 'Principles and Politics: Beyond the Impasse over the Embryo', in B. Waters and R. Cole-Turner, eds, *God and the Embryo: Religious Voices on Stem Cells and Cloning* (Washington, DC: Georgetown University Press, 2003), 88–97.

——, 'Soma, Psyche, Sin, and Salvation: Exploring the Relationship between Genetics and Theology', in M. L. Y. Chan and R. Chia, eds, *Beyond Determinism and Reductionism* (Adelaide: ATF Press, 2003), 16–35.

Collin, R., 'The Curious Case of an Accidental Artist', *Telegraph* (1 August 2011) <http://www.telegraph.co.uk/health/healthnews/8670516/The-curious-case-of-an-accidental-artist.html> accessed 22 June 2013.

Collins, F., 'Has the Revolution Arrived?', *Nature* 464/7289 (2010), 674–5.

——, 'Medical and Societal Consequences of the Human Genome Project', *New England Journal of Medicine* 341/1 (1999), 28–37.

Congregation for the Doctrine of the Faith, '*Donum Vitae*, Instruction on Respect for Human Life in Its Origin and on the Dignity of Procreation: Replies to Certain Questions of the Day', (1987) <http://www.vatican.va/roman_curia/congregations/cfaith/documents/rc_con_cfaith_doc_19870222_respect-for-human-life_en.html> accessed 28 March, 2013.

——, 'Instruction *Dignitas Personae* on Certain Bioethical Questions', (2008) <http://www.vatican.va/roman_curia/congregations/cfaith/documents/rc_con_cfaith_doc_20081208_dignitas-personae_en.html> accessed 4 April 2013.

Cooper, J. W., *Body, Soul and Life Everlasting: Biblical Anthropology and the Monism-Dualism Debate* (Grand Rapids, MI: Eerdmans, 1989).

Corbyn, Zoë, 'Misconduct Is the Main Cause of Life-Sciences Retractions', *Nature* 490 (2012), 21.

Coutelle, Charles, Carolyn Williams, Alan Handyside, Kate Hardy, Robert Winston, and Robert Williamson, 'Genetic Analysis of DNA from Single Human Oocytes: A Model for Preimplantation Diagnosis of Cystic Fibrosis', *BMJ* 299/6690 (1989), 22.

Crick, Francis, *The Astonishing Hypothesis* (London: Touchstone Books, 1994).

Crockett, M. J., L. Clark, M. D. Hauser, and T. W. Robbins, 'Serotonin Selectively Influences Moral Judgment and Behavior through Effects on Harm Aversion', *Proceedings of the National Academy of Sciences* 107/40 (2010), 17433–8.

Crowell, S. E, T. P. Beauchaine, E. McCauley, C. J. Smith, C. A. Vasilev, and A. L. Stevens, 'Parent–Child Interactions, Peripheral Serotonin, and Self-Inflicted Injury in Adolescents', *Journal of Consulting and Clinical Psychology* 76/1 (2008), 15–21.

Cullmann, Oscar, 'Immortality of the Soul or Resurrection of the Dead? The Witness of the New Testament', in Terence Penelhum, ed., *Immortality* (Belmont, CA: Wadsworth, 1973), 53–85.

Curley, J. P., C. L. Jensen, R. Mashoodh, and F. A. Champagne, 'Social Influences on Neurobiology and Behavior: Epigenetic Effects During Development', *Psychoneuroendocrinology* 36/3 (2011), 352–71.

Cyranoski, D, 'Retraction Record Rocks Community', *Nature* 489 (2012), 346–7.

Daily Mail Reporter, 'Ex-Street Fighter, 60, Turned into a Fanatical Artist by a Brain Haemorrhage That Physically Altered His Mind', *Daily Mail* (16 March 2010) <http://www.dailymail.co.uk/news/article-1256967/Man-survived-brain-haemorrhage-transformed-fanatical-artist-paints-18-hours-day.html> accessed 22 June 2013.

Daley, G. Q., 'Stem Cells: Roadmap to the Clinic', *The Journal of Clinical Investigation* 120/1 (2010), 8–10.

Davis, P. B., 'Cystic Fibrosis since 1938', *American Journal of Respiratory and Critical Care Medicine* 173/5 (2006), 475–82.

Dawkins, Richard, *The God Delusion* (London: Bantam Press, 2006).

De Dreu, C. K. W., L. L. Greer, G. A. Van Kleef, S. Shalvi, and M. J. J. Handgraaf, 'Oxytocin Promotes Human Ethnocentrism', *Proceedings of the National Academy of Sciences* 108/4 (2011), 1262–6.

de Grey, Aubrey D. N. J., ed. *Strategies for Engineered Negligible Senescence: Why Genuine Control of Aging May Be Foreseeable*. Vol. 1019, Annals of the New York Academy of Sciences (New York Academy of Sciences, 2004).

de Grey, Aubrey D. N. J., Bruce N. Ames, Julie K. Andersen, Andrzej Bartke, Judith Campisi, Christopher B. Heward, Roger J. M. McCarter, and Gregory Stock, 'Time to Talk SENS: Critiquing the Immutability of Human Aging', *Annals of the New York Academy of Sciences* 959/1 (2002), 452–62.

De Neve, Jan-Emmanuel, Slava Mikhaylov, Christopher T. Dawes, Nicholas A. Christakis, and James H. Fowler, 'Born to Lead? A Twin Design and Genetic Association Study of Leadership Role Occupancy', *The Leadership Quarterly* 24/1 (2013), 45–60.

Deane-Drummond, Celia, 'Bodies in Glass: A Virtue Approach to Ethical Quandaries in a Cyborg Age through a Recovery of Practical Wisdom', in D. Gareth Jones and R. John Elford, eds, *A Glass Darkly: Medicine and Theology in Further Dialogue* (Bern: Peter Lang, 2010), 61–79.

——, *Christ and Evolution: Wonder and Wisdom* (Minneapolis, MN: Fortress Press, 2009).

——, 'Future Perfect? God, the Transhuman Future and the Quest for Immortality', in C. Deane-Drummond and P. Scott, eds, *Future Perfect? God, Medicine and Human Identity* (London: T&T Clark, 2006), 168–82.

——, *Genetics and Christian Ethics* (Cambridge: Cambridge University Press, 2006).

——, 'How Might a Virtue Ethic Frame Debates in Human Genetics?', in Celia Deane-Drummond, ed., *Brave New World: Theology, Ethics and the Human Genome* (London: T&T Clark International, 2003), 225–52.

Department of Health, *The Removal, Retention and Use of Human Organs and Tissue from Postmortem Examination* (London: Her Majesty's Stationary Office, 2001).

Desmond, Jane, 'Postmortem Exhibitions: Taxidermied Animals and Plastinated Corpses in the Theaters of the Dead', *Configurations* 16/3 (2010), 347–78.

Dockery, C. A., R. Hueckel-Weng, N. Birbaumer, and C. Plewnia, 'Enhancement of Planning Ability by Transcranial Direct Current Stimulation', *The Journal of Neuroscience* 29/22 (2009), 7271–7.

Dodd, M. L., K. J. Klos, J. H. Bower, Y. E. Geda, K. A. Josephs, and J. E. Ahlskog, 'Pathological Gambling Caused by Drugs Used to Treat Parkinson Disease', *Archives of Neurology* 62/9 (2005), 1377–81.

Doidge, Norman, *The Brain That Changes Itself* (Melbourne: Scribe, 2010).

Dolan, S. M., 'Prenatal Genetic Testing', *Pediatric Annals* 38/8 (2009), 426–30.

Dolce, Linus, 'Injustice Perpetrated on the Dead', *The National Catholic Bioethics Quarterly* 10/4 (2010), 667–76.

Douglas, Thomas, 'Moral Enhancement Via Direct Emotion Modulation: A Reply to John Harris', *Bioethics* 27/3 (2013), 1601–68.

Dunstan, G. R., 'Review of Oliver O'Donovan. Begotten or Made?', *Religious Studies* 21/3 (1985), 415–16.

Eccles, J. C., *The Human Mystery* (Berlin: Springer, 1979).

——, *The Human Psyche* (Berlin: Springer, 1980).

Eccleston, Alex, Natalie DeWitt, Chris Gunter, Barbara Marte, and Deepa Nath, 'Epigenetics', *Nature* 447/7143 (2007), 395.

Edwards, Robert, *Life before Birth* (London: Hutchinson, 1989).

Elford, R. John, and D. Gareth Jones, eds, *A Tangled Web: Medicine and Theology in Dialogue* (Bern: Peter Lang, 2009).

Elliott, R., B. J. Sahakian, K. Matthews, A. Bannerjea, J. Rimmer, and T. W. Robbins, 'Effects of Methylphenidate on Spatial Working Memory and Planning in Healthy Young Adults', *Psychopharmacology* 131/2 (1997), 196–206.

European Society of Human Reproduction and Embryology, 'The World's Number of IVF and ICSI Babies Has Now Reached a Calculated Total of 5 Million', (2 July 2012) <http://www.eshre.eu/ESHRE/English/Press-Room/Press-Releases/Press-releases-2012/5-million-babies/page.aspx/1606> accessed 18 June 2013.

Feinberg, A. P., 'Phenotypic Plasticity and the Epigenetics of Human Disease', *Nature* 447/7143 (2007), 433–40.

Fenton, A., and S. Alpert, 'Extending Our View on Using BCIs for Locked-in Syndrome', *Neuroethics* 1 (2008), 119–32.

Feresin, Emiliano, 'Lighter Sentence for Murderer with "Bad Genes"', *Nature* 10 (2009), 1038.

Ferraretti, A. P., V. Goossens, J. de Mouzon, S. Bhattacharya, J. A. Castilla, V. Korsak, M. Kupka, K. G. Nygren, and A. Nyboe Andersen, 'Assisted Reproductive Technology in Europe, 2008: Results Generated from European Registers by ESHRE', *Human Reproduction* 27/9 (2012), 2571–84.

Fisher, Paul J., Dragana Turic, Nigel M. Williams, Peter McGuffin, Philip Asherson, David Ball, Ian Craig, Thalia Eley, Linzy Hill, and Karen Chorney, 'DNA Pooling Identifies QTLs on Chromosome 4 for General Cognitive Ability in Children', *Human Molecular Genetics* 8/5 (1999), 915–22.

Fitz, Nicholas S., and Peter B. Reiner, 'The Challenge of Crafting Policy for Do-It-Yourself Brain Stimulation', *Journal of Medical Ethics* (2013), Advance online publication, 3 June 2013, doi: 10.1136/medethics-2013-101458.

Fong, A. J., R. R. Roy, R. M. Ichiyama, I. Lavrov, G. Courtine, Y. Gerasimenko, Y. C. Tai, J. Burdick, and V. R. Edgerton, 'Recovery of Control of Posture and Locomotion after a Spinal Cord Injury: Solutions Staring Us in the Face', *Progress in Brain Research* 175 (2009), 393–418.

Fox, D., 'Brain Buzz', *Nature* 472/7342 (2011), 156–8.

Fragouli, E., 'Preimplantation Genetic Diagnosis: Present and Future', *Journal of Assisted Reproduction and Genetics* 24/6 (2007), 201–7.

Fukuyama, F., *Our Posthuman Future: Consequences of the Biotechnology Revolution* (New York, NY: Picador, 2002).

Fusco, S., C. Ripoli, M. V. Podda, S. C. Ranieri, L. Leone, G. Toietta, M. W. McBurney, G. Schütz, A. Riccio, C. Grassi, T. Galeotti, and G. Pani, 'A Role for Neuronal cAMP Responsive-Element Binding (CREB)-1 in Brain Responses to Calorie Restriction', *Proceedings of the National Academy of Sciences* 109/2 (2012), 621–6.

Garreau, J, *Radical Evolution: The Promise and Peril of Enhancing Our Minds, Our Bodies – and What It Means to Be Human* (New York, NY: Doubleday, 2005).

Gibbons, A., 'American Association of Physical Anthropologists Meeting. Tracking the Evolutionary History of a "Warrior" Gene', *Science* 304/5672 (2004), 818.

Gillett, G. R., 'Cyborgs and Moral Identity', *Journal of Medical Ethics* 32/2 (2006), 79–83.

Gilman, S. L., *Creating Beauty to Cure the Soul* (Durham: Duke University Press, 1998).

Glannon, Walter, 'Decelerating and Arresting Human Aging', *Medical Enhancement and Posthumanity* (2008), 175–89.

——, 'Psychopharmacology and Memory', *Journal of Medical Ethics* 32 (2006), 74–8.

——, 'Stimulating Brains, Altering Minds', *Journal of Medical Ethics* 35/5 (2009), 289–392.

Goddard, Andrew, 'The Place of the Bible in Medical Ethics', in D. Gareth Jones and R. John Elford, eds, *A Glass Darkly: Medicine and Theology in Further Dialogue* (Bern: Peter Lang, 2010), 133–56.

Graham, Elaine L., *Representations of the Post/Human: Monsters, Aliens and Others in Popular Culture* (Manchester: Manchester University Press, 2002).

Greely, Henry, Barbara Sahakian, John Harris, Ronald C. Kessler, Michael Gazzaniga, Philip Campbell, and Martha J. Farah, 'Towards Responsible Use of Cognitive-Enhancing Drugs by the Healthy', *Nature* 456 (2008), 702–5.

Green, Joel B, *Body, Soul, and Human Life: The Nature of Humanity in the Bible* (Grand Rapids MI: Baker Academic, 2008).

Green, Joel B., and Stuart L. Palmer, eds, *In Search of the Soul* (Downers Grove, IL: Inter-Varsity Press, 2005).

Grön, Georg, Matthias Kirstein, Axel Thielscher, Matthias W. Riepe, and Manfred Spitzer, 'Cholinergic Enhancement of Episodic Memory in Healthy Young Adults', *Psychopharmacology* 182/1 (2005), 170–9.

Guttmacher, A. E., and F. S. Collins, 'Genomic Medicine – a Primer', *The New England Journal of Medicine* 347/19 (2002), 1512–20.

Hagerty, Barbara Bradley, 'A Neuroscientist Uncovers a Dark Secret', *National Public Radio* (29 June 2010) <http://www.npr.org/templates/story/story.php?storyId=127888976> accessed 5 April 2013.

Hall, W. D., R. Mathews, and K. I. Morley, 'Being More Realistic About the Public Health Impact of Genomic Medicine', *PLoS Medicine* 7/10 (2010), e1000347.

Hamer, Dean H., *The God Gene: How Faith Is Hardwired into Our Brains* (New York: Doubleday, 2004).

Hamer, Dean H., Stella Hu, Victoria L. Magnuson, Nan Hu, and Angela M. Pattatucci, 'A Linkage between DNA Markers on the X Chromosome and Male Sexual Orientation', *Science* 261/5119 (1993), 321–7.

Haraway, Donna, 'A Cyborg Manifesto: Science, Technology and Socialist Feminism in the Late Twentieth Century', in Donna Haraway, ed., *Simeans, Cyborgs and Women* (London: Free Association Books, 1991), 149–81.

——, 'Cyborg to Companion Species: Reconfiguring Kinship in Technoscience', in Donna Haraway, ed., *The Haraway Reader* (London: Routledge, 2004), 295–320.

——, *Simians, Cyborgs, and Women: The Reinvention of Nature* (London: Free Association Books, 1991).

Harbaugh, W. T., U. Mayr, and D. R. Burghart, 'Neural Responses to Taxation and Voluntary Giving Reveal Motives for Charitable Donations', *Science* 316/5831 (2007), 1622–5.

Harris, John, 'Cloning and Human Dignity', *Cambridge Quarterly of Healthcare Ethics* 7/2 (1998), 163–7.

——, 'Moral Enhancement and Freedom', *Bioethics* 25/2 (2011), 102–11.

Hasker, William, *The Emergent Self* (Ithaca, NY: Cornell University Press, 1999).

Hauskeller, Michael, 'The Moral Status of Post-Persons', *Journal of Medical Ethics* 39/2 (2013), 76–7.

Hayashi, K., S. Ogushi, K. Kurimoto, S. Shimamoto, H. Ohta, and M. Saitou, 'Offspring from Oocytes Derived from In Vitro Primordial Germ Cell-Like Cells in Mice', *Science* 338/6109 (2012), 971–5.

Hayashi, K., H. Ohta, K. Kurimoto, S. Aramaki, and M. Saitou, 'Reconstitution of the Mouse Germ Cell Specification Pathway in Culture by Pluripotent Stem Cells', *Cell* 146/4 (2011), 519–32.

Haynes, J. D., and G. Rees, 'Decoding Mental States from Brain Activity in Humans', *Nature Reviews Neuroscience* 7/7 (2006), 523–34.

Hays, R. B., *The Moral Vision of the New Testament: Community, Cross, New Creation: A Contemporary Introduction to New Testament Ethics* (San Francisco: HarperSanFrancisco, 1996).

Heard, A, 'Technology Makes Us Optimistic; They Want to Live', *The New York Times Magazine* (28 September 1997) 84–9. <http://www.nytimes.com/1997/09/28/magazine/technology-makes-us-optimistic-they-want-to-live.html?pagewanted=all&src=pm> accessed 11 April 2013.

Hefner, Philip, *Technology and Human Becoming* (Minneapolis, MN: Fortress Press, 2003).

Heinz, A., R. Kipke, H. Heimann, and U. Wiesing, 'Cognitive Neuroenhancement: False Assumptions in the Ethical Debate', *Journal of Medical Ethics* 38/6 (2012), 372–5.

Heinz, Andreas, Anne Beck, Sabine M. Grüsser, Anthony A. Grace, and Jana Wrase, 'Identifying the Neural Circuitry of Alcohol Craving and Relapse Vulnerability', *Addiction Biology* 14/1 (2009), 108–18.

Hochberg, L. R., M. D. Serruya, G. M. Friehs, J. A. Mukand, M. Saleh, A. H. Caplan, A. Branner, D. Chen, R. D. Penn, and J. P. Donoghue, 'Neuronal Ensemble Control of Prosthetic Devices by a Human with Tetraplegia', *Nature* 442/7099 (2006), 164–71.

Hoffman, J. L., 'Clinical Observations Concerning Schizophrenic Patients Treated by Prefrontal Leukotomy', *New England Journal of Medicine* 241 (1949), 233–6.

Honigman, R., and D. J. Castle, 'Aging and Cosmetic Enhancement', *Clinical Interventions in Aging* 1/2 (2006), 115–19.

Hu, Stella, Angela M. L. Pattatucci, Lin Li Chavis Patterson, David W. Fulker, Stacey S. Cherny, Leonid Kruglyak, and Dean H. Hamer, 'Linkage between Sexual Orientation and Chromosome Xq28 in Males but Not in Females', *Nature Genetics* 11/3 (1995), 248–56.

Hubbard, Ruth, and Elijah Wald, *Exploding the Gene Myth: How Genetic Information Is Produced and Manipulated by Scientists, Physicians, Employers, Insurance Companies, Educators, and Law Enforcers* (Boston, MA: Beacon Press, 1993).

Hui, E. C., *At the Beginning of Life: Dilemmas in Theological Bioethics* (Downers Grove, IL: Inter Varsity Press, 2002).

'Human Fertilisation and Embryology Act', (2008) <http://www.legislation.gov.uk/ukpga/2008/22/contents> accessed 18 June 2013.

Human Fertilisation and Embryology Authority, 'PGD Conditions Licensed by the HFEA', <http://www.hfea.gov.uk/cps/hfea/gen/pgd-screening.htm> accessed 18 June 2013.

Hummer, Robert A., and Juanita J. Chinn, 'Race/Ethnicity and US Adult Mortality', *Du Bois Review: Social Science Research on Race* 8/1 (2011), 5–24.

Huxley, Aldous, *Brave New World*, 1958 ed. (Harmondsworth: Penguin Books, 1932).

'Interview with Robert Edwards', *The Times* (24 July 2003).

Ip, King-Tak, ed. *The Bioethics of Regenerative Medicine* (Netherlands: Springer, 2009).

——, 'Introduction: Regenerative Medicine at the Heart of the Culture Wars', in King-Tak Ip, ed., *The Bioethics of Regenerative Medicine* (Netherlands: Springer, 2009), 3–10.

Jackson, Jennifer, 'Unproven Treatment in Childhood Oncology – How Far Should Paediatricians Co-operate?', *Journal of Medical Ethics* 20/2 (1994), 77–9.

Jeeves, Malcolm, 'Human Nature: An Integrated Picture', in Joel B. Green, ed., *What About the Soul?: Neuroscience and Christian Anthropology* (Nashville, TN: Abingdon, 2004), 171–89.

Jespersen, T. Christine, and Alicita Rodríguez, 'Forced Impregnation and Masculinist Utopia', in T. Christine Jespersen, Alicita Rodríguez and Joseph Starr, eds, *The Anatomy of Body Worlds: Critical Essays on the Plastinated Cadavers of Gunther Von Hagens* (Jefferson, NC: McFarland & Co, 2009), 166–75.

Jha, A., 'Baby Mice Created from Stem Cells', *The Guardian* (4 October 2012) <http://www.guardian.co.uk/science/2012/oct/04/baby-mice-stem-cells> accessed 24 June 2013.

Johnston, E., 'Nature Versus Nurture: Are We Missing a Third Option?', (Thesis, Master of Health Sciences, University of Otago, 2013).

Jones, D., 'Moral Psychology: The Depths of Disgust', *Nature* 447/7146 (2007), 768–71.

Jones, D. A., *The Soul of the Embryo: An Inquiry into the Status of the Human Embryo in Christian Tradition* (London: Continuum, 2004).

Jones, D. Gareth, 'Abortion: An Alternative to the Conflict Paradigm', in B. Richards and V. Pfitzner, eds, *Issues at the Borders of Life* (Adelaide: ATF Press, 2010), 11–21.

——, *Bioethics: When the Challenges of Life Become Too Difficult* (Adelaide: ATF Press, 2007).

——, 'The Biomedical Technologies: Prospects and Challenges', in D. Gareth Jones and R. John Elford, eds, *A Glass Darkly: Medicine and Theology in Further Dialogue* (Bern: Peter Lang, 2010), 9–32.

——, *Brave New People: Ethical Issues at the Commencement of Life* (Leicester: Inter-Varsity Press, 1984).

——, 'Conclusion: The Necessity of Dialogue', in D. Gareth Jones and R. John Elford, eds, *A Glass Darkly: Medicine and Theology in Further Dialogue* (Bern: Peter Lang, 2010), 211–38.

——, *Designers of the Future: Who Should Make the Decisions?* (Oxford: Monarch, 2005).

——, 'Enhancement: Are Ethicists Excessively Influenced by Baseless Speculations?', *Medical Humanities* 32/2 (2006), 77–81.

——, 'Enhancement: Is Baseless Speculation Misleading Theologians and Bioethicists?', in R. J. Elford and D. G. Jones, eds, *A Tangled Web: Medicine and Theology in Dialogue* (Bern: Peter Lang, 2009), 123–42.

——, 'The Human Body: An Anatomist's Journey from Death to Life', in R. J. Elford and D. G. Jones, eds, *A Tangled Web: Medicine and Theology in Dialogue* (Bern: Peter Lang, 2009), 105–21.

——, 'The Human Cadaver: An Assessment of the Value We Place on the Dead Body', *Perspectives on Science and Christian Faith* 47 (1995), 43–51.

——, 'The Importance of Realism in Assessing Technological Possibilties: The Role of Christian Thinking', *Christian Perspectives on Science and Technology, ISCAST Online Journal* 9 (June 2013) <http://www.iscast.org/journal/articles/Jones_G_2013-06_Realism> accessed 24 June 2013.

——, *Manufacturing Humans: The Challenge of the New Reproductive Technologies* (Leicester, England: Inter-Varsity Press, 1987).

——, 'Moral Enhancement as a Technological Imperative', *Perspectives on Science and Christian Faith* 65/3 (2013), 187–95.

——, 'A Neurobiological Portrait of the Human Person: Finding a Context for Approaching the Brain', in Joel B. Green, ed., *What About the Soul? Neuroscience and Christian Anthropology* (Nashville, TN: Abingdon Press, 2004), 31–46.

——, *Our Fragile Brains: A Christian Perspective on Brain Research* (Downers Grove, IL: Inter-Varsity Press, 1980).

——, 'Responses to the Human Embryo and Embryonic Stem Cells: Scientific and Theological Assessments', *Science & Christian Belief* 17/2 (2005), 199–221.

——, *Valuing People: Human Value in a World of Medical Technology* (Carlisle: Paternoster Press, 1999).

Jones, D. Gareth, and R. John Elford, *A Glass Darkly: Medicine and Theology in Further Dialogue* (Bern: Peter Lang, 2010).

Jones, D. Gareth, and Maja I. Whitaker, 'Scientific Fraud: The Demise of Idealistic Science', in R. J. Elford and D. G. Jones, eds, *A Tangled Web: Medicine and Theology in Dialogue* (Bern: Peter Lang, 2009), 89–104.

——, *Speaking for the Dead: The Human Body in Biology and Medicine*, 2nd edn (Aldershot: Ashgate, 2009).

——, 'Transforming the Human Body', in C. Blake, C. Molloy and S. Shakespeare, eds, *Beyond Human: From Animality to Transhumanism* (London: Continuum, 2012), 254–79.

Juengst, E. T., R. H. Binstock, M. J. Mehlman, and S. G. Post, 'Antiaging Research and the Need for Public Dialogue', *Science* 299/5611 (2003), 1323.

Juengst, E. T., R. H. Binstock, M. Mehlman, S. G. Post, and P. Whitehouse, 'Biogerontology, "Anti-Aging Medicine," and the Challenges of Human Enhancement', *The Hastings Center Report* 33/4 (2003), 21–30.

Jung, Hyun Ho, C.-H. Kim, Jong Hee Chang, Yong Gou Park, Sang Sup Chung, and Jin Woo Chang, 'Bilateral Anterior Cingulotomy for Refractory Obsessive-Compulsive Disorder: Long-Term Follow-up Results', *Stereotactic and Functional Neurosurgery* 84/4 (2006), 184–9.

Junker-Kenny, M., 'Genetic Perfection, or Fulfilment of Creation in Christ?', in C. Deane-Drummond and P. Scott, eds, *Future Perfect? God, Medicine and Human Identity* (London: T&T Clark, 2006), 155–67.

Kadosh, Roi Cohen, Neil Levy, Jacinta O'Shea, Nicholas Shea, and Julian Savulescu, 'The Neuroethics of Non-Invasive Brain Stimulation', *Current Biology* 22/4 (2012), R108–11.

Kahn, Axel, 'Clone Mammals ... Clone Man?', *Nature* 386 (1997), 119.

Kaiser, L. R., 'The Future of Multihospital Systems', *Topics in Health Care Financing* 18/4 (1992), 32–45.

Kalaria, R. N., G. E. Maestre, R. Arizaga, R. P. Friedland, D. Galasko, K. Hall, J. A. Luchsinger, A. Ogunniyi, E. K. Perry, F. Potocnik, M. Prince, R. Stewart, A. Wimo, Z. X. Zhang, and P. Antuono, 'Alzheimer's Disease and Vascular Dementia in Developing Countries: Prevalence, Management, and Risk Factors', *Lancet Neurol* 7/9 (2008), 812–26.

Kass, Leon R, 'Triumph or Tragedy – the Moral Meaning of Genetic Technology', *American Journal of Jurisprudence* 45 (2000), 1–16.

Kemnitz, J. W., 'Calorie Restriction and Aging in Nonhuman Primates', *ILAR Journal* 52/1 (2011), 66–77.

Kennedy, D. P., J. Glascher, J. M. Tyszka, and R. Adolphs, 'Personal Space Regulation by the Human Amygdala', *Nature Neuroscience* 12/10 (2009), 1226–7.

Kepler, Johannes, 'Letter to Michael Maestlin (3 Oct 1595)', in, *Johannes Kepler Gesammelte Werke* (1937).

Khoshbin, L. S., and S. Khoshbin, 'Imaging the Mind, Minding the Image: An Historical Introduction to Brain Imaging and the Law', *American Journal of Law & Medicine* 33/2–3 (2007), 171–92.

Kim-Cohen, Julia, Avshalom Caspi, Alan Taylor, Benjamin Williams, Rhiannon Newcombe, Ian W. Craig, and Terrie E. Moffitt, 'MAOA, Maltreatment, and Gene–Environment Interaction Predicting Children's Mental Health: New Evidence and a Meta-Analysis', *Molecular Psychiatry* 11/10 (2006), 903–13.

King, D. S., 'Preimplantation Genetic Diagnosis and the "New" Eugenics', *Journal of Medical Ethics* 25/2 (1999), 176–82.

King, Michael, Maja Whitaker, and D. Gareth Jones, 'Speculative Ethics: Valid Enterprise or Tragic Cul-De-Sac?', in Abraham Rudnick, ed., *Bioethics in the 21st Century* (InTech, 2011), 139–58.

King, Mike R., Maja I. Whitaker, and D. Gareth Jones, 'I See Dead People: Insights from the Humanities into the Nature of Plastinated Cadavers', *Journal of Medical Humanities* (2013), Advance online publication, 22 March 2013, doi: 10.1007/s10912-013-9230-z.

Kirkpatrick, C. J., S. Fuchs, K. Peters, C. Brochhausen, M. I. Hermanns, and R. E. Unger, 'Visions for Regenerative Medicine: Interface between Scientific Fact and Science Fiction', *Artificial Organs* 30/10 (2006), 822–7.

Koenigs, M., L. Young, R. Adolphs, D. Tranel, F. Cushman, M. Hauser, and A. Damasio, 'Damage to the Prefrontal Cortex Increases Utilitarian Moral Judgements', *Nature* 446/7138 (2007), 908–11.

Kosfeld, M., M. Heinrichs, P. J. Zak, U. Fischbacher, and E. Fehr, 'Oxytocin Increases Trust in Humans', *Nature* 435/7042 (2005), 673–6.

Koubova, J., and L. Guarente, 'How Does Calorie Restriction Work?', *Genes & Development* 17/3 (2003), 313–21.

Kramer, P. D., *Listening to Prozac* (New York: Viking, 1993).

Kuhn, J., T. O. Grundler, D. Lenartz, V. Sturm, J. Klosterkotter, and W. Huff, 'Deep Brain Stimulation for Psychiatric Disorders', *Deutsches Arzteblatt International* 107/7 (2010), 105–13.

Kurzweil, R., 'Human Body Version 2.0', (16 February 2003) <http://www.kurzweilai.net/human-body-version-20> accessed 11 April 2013.

——, 'Reinventing Humanity: The Future of Human-Machine Intelligence', *The Futurist* 40/2 (2006), 39–46.

——, *The Singularity Is Near* (New York, NY: Viking, 2005).

Kutz, Gregory, D., 'Nutrigenetic Testing: Tests Purchased from Four Web Sites Mislead Consumers', Government Accountability Office (27 July 2006) <http://www.gao.gov/products/GAO-06-977T> accessed 18 June 2013.

Kwok, R., 'Neuroprosthetics: Once More, with Feeling', *Nature* 497/7448 (2013), 176–8.

Lamb, N., 'The Genetics of Eye Color', *Community Outreach* (Fall 2009) <http://www.hudsonalpha.org/sites/default/files/pdf/genetics_of_eye_color.pdf> accessed 18 June 2013.

Lampe, Peter, 'Paul's Concept of a Spiritual Body', in Ted Peters, Robert John Russell and Michael Welker, eds, *Resurrection: Theological and Scientific Assessments* (Grand Rapids, MI: William B. Eerdmans, 2002), 103–14.

Lane, M., D. Ingram, and G. Roth, 'The Serious Search for an Antiaging Pill', *Scientific American* 287/2 (2004), 36–41.

Lantos, J. D., ed. *Controversial Bodies: Thoughts on the Public Display of Plastinated Corpses* (Baltimore, MD: Johns Hopkins University Press, 2011).

Lee, S. E., S. B. Simons, S. A. Heldt, M. Zhao, J. P. Schroeder, C. P. Vellano, D. P. Cowan, S. Ramineni, C. K. Yates, and Y. Feng, 'RGS14 Is a Natural Suppressor of Both Synaptic Plasticity in CA2 Neurons and Hippocampal-Based Learning and Memory', *Proceedings of the National Academy of Sciences* 107/39 (2010), 16994–8.

Levy, Neil, 'Ecological Engineering: Reshaping Our Environments to Achieve Our Goals', *Philosophy & Technology* 25/4 (2012), 589–604.

Lewin, Tamar, 'College Bound, DNA Swab in Hand', *New York Times* (18 May 2010).

Lewontin, R. C., Steven Rose, and Leon J. Kamin, *Not in Our Genes: Biology, Ideology and Human Nature* (New York, NY: Pantheon Books, 1984).

Liggins Institute, 'Developmental Epigenetics', <http://www.liggins.auckland.ac.nz/uoa/home/about/research-themes/developmentalepigenetics> accessed 18 June 2013.

Ling, John R., *When Does Human Life Begin?* (Newcastle upon Tyne: The Christian Institute, 2011).

Lobo, I., 'Environmental Influences on Gene Expression', *Nature Education* 1 (2008), 1.

Looy, Heather, 'Psychology at the Theological Frontiers', *Perspectives on Science and Christian Faith* 65/3 (2013), 147–55.

Lynch, Gary, and Christine M. Gall, 'Ampakines and the Threefold Path to Cognitive Enhancement', *Trends in Neurosciences* 29/10 (2006), 554–62.

Lysaght, T., and A. V. Campbell, 'The Ethics of Regenerative Medicine: Broadening the Scope Beyond the Moral Status of Embryos', in Akira Akabayashi, ed., *The Future of Bioethics: International Dialogues* (Oxford: Oxford University Press, 2014), 5–26.

Macchiarini, P., P. Jungebluth, T. Go, M. Asnaghi, L. E. Rees, T. A. Cogan, A. Dodson, J. Martorell, S. Bellini, and P. P. Parnigotto, 'Clinical Transplantation of a Tissue-Engineered Airway', *Lancet* 372/9655 (2008), 2023–30.

MacKay, Donald M., 'Brain Science and the Soul', in R. L. Gregory, ed., *The Oxford Companion to the Mind* (Oxford: Oxford University Press, 1987), 723–5.

——, *Brains, Machines and Persons* (London: Collins, 1980).

——, *Human Science and Human Dignity* (London: Hodder and Stoughton, 1979).

——, *The Open Mind and Other Essays*. Edited by M. Tinker (Leicester: Inter-Varsity Press, 1988).

MacKellar, Calum, and David A. Jones, eds, *Chimera's Children: Ethical, Philosophical and Religious Perspectives on Human–Nonhuman Combinations* (London: Continuum, 2012).

Macklon, N. S., J. P. Geraedts, and B. C. Fauser, 'Conception to Ongoing Pregnancy: The "Black Box" of Early Pregnancy Loss', *Human Reproduction Update* 8/4 (2002), 333–43.

Malek, J., 'Deciding against Disability: Does the Use of Reproductive Genetic Technologies Express Disvalue for People with Disabilities?', *Journal of Medical Ethics* 36/4 (2010), 217–21.

Mandler, George, 'Apart from Genetics: What Makes Monozygotic Twins Similar?', *Journal of Mind and Behavior* 22/2 (2001), 147–59.

Manipalviratn, Somjate, Alan DeCherney, and James Segars, 'Imprinting Disorders and Assisted Reproductive Technology', *Fertility and Sterility* 91/2 (2009), 305–15.

Markestad, T., P. I. Kaaresen, A. Ronnestad, H. Reigstad, K. Lossius, S. Medbo, G. Zanussi, I. E. Engelund, R. Skjaerven, and L. M. Irgens, 'Early Death, Morbidity, and Need of Treatment among Extremely Premature Infants', *Pediatrics* 115/5 (2005), 1289–98.

Markham, J. A., and W. T. Greenough, 'Experience-Driven Brain Plasticity: Beyond the Synapse', *Neuron Glia Biology* 1/4 (2004), 351–63.

Marsh, Abigail A., Samantha L. Crowe, H. Yu Henry, Elena K. Gorodetsky, David Goldman, and R. J. R. Blair, 'Serotonin Transporter Genotype (5-HTTLPR) Predicts Utilitarian Moral Judgments', *PloS One* 6/10 (2011), e25148.

Matthews, Keith, and Muftah S. Eljamel, 'Status of Neurosurgery for Mental Disorder in Scotland: Selective Literature Review and Overview of Current Clinical Activity', *The British Journal of Psychiatry* 182/5 (2003), 404–11.

May, W. F., 'Religious Justification for Donating Body Parts', *Hastings Center Report* 15/1 (1985), 38–42.

McCarthy, B., *Fertility & Faith: The Ethics of Human Fertilization* (Leicester: Inter-Varsity Press, 1997).

McCay, C. M., M. F. Crowell, and L. A. Maynard, 'The Effect of Retarded Growth Upon the Length of Life Span and Upon the Ultimate Body Size', *Journal of Nutrition* 10/1 (1935), 63–79.

McColl, K. E., H. Watabe, and M. H. Derakhshan, 'Sporadic Gastric Cancer: A Complex Interaction of Genetic and Environmental Risk Factors', *American Journal of Gastroenterology* 102/9 (2007), 1893–5.

McConnel, C., and L. Turner, 'Medicine, Ageing and Human Longevity', *EMBO Reports* 6/S1 (2005), S59–62.

McGraw, J. J., 'Tongues of Men and Angels: Assessing the Neural Correlates of Glossolalia', in D. Cave and R. S. Norris, eds, *The Body and Religion: Modern Science and the Construction of Religious Meaning* (Leiden: Brill, 2012), 57–79.

McKenny, Gerald P., 'The Ethics of Regenerative Medicine: Beyond Humanism and Posthumanism', in King-Tak Ip, ed., *The Bioethics of Regenerative Medicine* (Netherlands: Springer, 2009), 155–69.

——, 'Technologies of Desire Theology, Ethics, and the Enhancement of Human Traits', *Theology Today* 59/1 (2002), 90–103.

McNamara, Patrick, 'The Motivational Origins of Religious Practices', *Zygon* 37 (2002), 143–60.

Mealey, Ann Marie, '*Dignitas Personae*', in D. Gareth Jones and R. John Elford, eds, *A Glass Darkly: Medicine and Theology in Further Dialogue* (Bern: Peter Lang, 2010), 111–29.

Mehta, M. A., A. M. Owen, B. J. Sahakian, N. Mavaddat, J. D. Pickard, and T. W. Robbins, 'Methylphenidate Enhances Working Memory by Modulating Discrete Frontal and Parietal Lobe Regions in the Human Brain', *Journal of Neuroscience* 20/6 (2000), RC65.

Meikle, J., 'Life Expectancy Rises in UK but North-South Divide Widens', *The Guardian* (2011) <http://www.guardian.co.uk/society/2011/jun/08/life-expectancy-north-south-divide-widens> accessed 26 June 2013.

Meilaender, Gilbert, *Should We Live Forever? The Ethical Ambiguities of Aging* (Grand Rapids, MI: Eerdmans, 2013).

Messer, Neil, 'Christian Engagement with Public Bioethics in Britain: The Case of Human Admixed Embryos', *Christian Bioethics* 15/1 (2009), 31–53.

——, *Respecting Life: Theology and Bioethics* (London: SCM Press, 2011).

Meyer, J. H., S. McMain, S. H. Kennedy, L. Korman, G. M. Brown, J. N. DaSilva, A. A. Wilson, T. Blak, R. Eynan-Harvey, V. S. Goulding, S. Houle, and P. Links, 'Dysfunctional Attitudes and 5-HT2 Receptors During Depression and Self-Harm', *American Journal of Psychiatry* 160/1 (2003), 90–9.

Midgely, M., 'Biotechnology and Monstrosity', *Hastings Center Report* 30 (2000), 7–15.

Miller, R. A., 'Extending Life: Scientific Prospects and Political Obstacles', *Milbank Quarterly* 80/1 (2002), 155–74.

Milner, Conan, 'Study Drugs Unsafe and Unethical, Say Neurologists', *Epoch Times*, (27 March 2013). <http://www.theepochtimes.com/n2/united-states/study-drugs-unsafe-and-unethical-say-neurologists-369451.html> accessed 1 April 2013.

Minzenberg, Michael J., and Cameron S. Carter, 'Modafinil: A Review of Neurochemical Actions and Effects on Cognition', *Neuropsychopharmacology* 33/7 (2007), 1477–502.

Mironov, V., R. P. Visconti, and R. R. Markwald, 'What Is Regenerative Medicine? Emergence of Applied Stem Cell and Developmental Biology', *Expert Opinion on Biological Therapy* 4/6 (2004), 773–81.

MIT Media Lab, 'Wearable Computing at the MIT Media Lab', (2005) <http://www.media.mit.edu/wearables/> accessed 11 April 2013.

Mohler, R. Albert, 'Christian Morality and Test Tube Babies, Part One', (29 September 2005) <http://www.albertmohler.com/2005/09/29/christian-morality-and-test-tube-babies-part-one-2/> accessed 17 June 2013.

——, 'Christian Morality and Test Tube Babies, Part Two', (12 May 2006) <http://www.albertmohler.com/2006/05/12/christian-morality-and-test-tube-babies-part-two/> accessed 17 June 2013.

Moll, J., R. de Oliveira-Souza, I. E. Bramati, and J. Grafman, 'Functional Networks in Emotional and Nonmoral Social Judgements', *NeuroImage* 26 (2002), 696–703.

Moll, J., R. de Oliveira-Souza, F. T. Moll, F. A. Ignacio, I. E. Bramati, E. M. Caparelli-Daquer, and P. J. Eslinger, 'The Moral Affiliations of Disgust: A Functional MRI Study', *Cognitive and Behavioral Neurology* 18/1 (2005), 68–78.

Moll, J., F. Krueger, R. Zahn, M. Pardini, R. de Oliveira-Souza, and J. Grafman, 'Human Fronto-Mesolimbic Networks Guide Decisions About Charitable Donation', *Proceedings of the National Academy of Sciences of the United States of America* 103/42 (2006), 15623–8.

Moreland, J. P., and S. B. Rae, *Body and Soul: Human Nature and the Crisis in Ethics* (Downers Grove, IL: Inter-Varsity Press, 2000).

Mpakopoulou, Maria, Haralambos Gatos, Alexandros Brotis, Konstantinos N. Paterakis, and Kostas N. Fountas, 'Stereotactic Amygdalotomy in the Management of Severe Aggressive Behavioral Disorders', *Neurosurgical Focus* 25/1 (2008), E6.

Mummery, C., D. Ward-van Oostwaard, P. Doevendans, R. Spijker, S. van den Brink, R. Hassink, M. van der Heyden, T. Opthof, M. Pera, A. B. de la Riviere, R. Passier, and L. Tertoolen, 'Differentiation of Human Embryonic Stem Cells to Cardiomyocytes: Role of Coculture with Visceral Endoderm-Like Cells', *Circulation* 107/21 (2003), 2733–40.

Murphy, Nancey, *Bodies and Souls, or Spirited Bodies?* (Cambridge, MA: Cambridge University Press, 2006).

Murphy, Nancey, and Warren S. Brown, *Did My Neurons Make Me Do It? Philosophical and Neurobiological Perspectives on Moral Responsibility and Free Will* (Oxford: Oxford University Press, 2007).

Nelkin, Dorothy, and M. Susan Lindee, *The DNA Mystique: The Gene as a Cultural Icon*, 2nd edn (Ann Arbor, MI: University of Michigan Press, 2004).

Neville, H., and D. Bavelier, 'Human Brain Plasticity: Evidence from Sensory Deprivation and Altered Language Experience', *Progress in Brain Research* 138 (2002), 177–88.

Newberg, A., A. Alavi, M. Baime, M. Pourdehnad, J. Santanna, and E. d'Aquili, 'The Measurement of Regional Cerebral Blood Flow During the Complex Cognitive Task of Meditation: A Preliminary SPECT Study', *Psychiatry Research: Neuroimaging* 106/2 (2001), 113–22.

Newberg, A. B., N. A. Wintering, D. Morgan, and M. R. Waldman, 'The Measurement of Regional Cerebral Blood Flow During Glossolalia: A Preliminary SPECT Study', *Psychiatry Research: Neuroimaging* 148/1 (2006), 67–71.

Newberg, A., M. Pourdehnad, A. Alavi, and E. G. d'Aquili, 'Cerebral Blood Flow During Meditative Prayer: Preliminary Findings and Methodological Issues', *Perceptual and Motor Skills* 97/2 (2003), 625–30.

Newman, H. H., F. N. Freeman, and K. J. Holzinger, 'Twins: A Study of Heredity and Environment', *Eugenics Review* 30/1 (1937), 61–2.

Nicholls, E. H., 'Selling Anatomy: The Role of the Soul', *Endeavour* 26/2 (2002), 47.

Nitsche, Michael A., Leonardo G. Cohen, Eric M. Wassermann, Alberto Priori, Nicolas Lang, Andrea Antal, Walter Paulus, Friedhelm Hummel, Paulo S. Boggio, and Felipe Fregni, 'Transcranial Direct Current Stimulation: State of the Art 2008', *Brain Stimulation* 1/3 (2008), 206–23.

Norman, R. J., D. Dewailly, R. S. Legro, and T. E. Hickey, 'Polycystic Ovary Syndrome', *Lancet* 370/9588 (2007), 685–97.

Northcott, M. S., 'Concept Art, Clones, and Co-Creators: The Theology of Making', *Modern Theology* 21/2 (2005), 219–36.

Nutt, A. E., *Shadows Bright as Glass: The Extraordinary Transformation of One Man's Brain and the Neuroscience That Makes Us Who We Are* (Piatkus Books, 2011).

O'Donovan, Oliver, *Begotten or Made?* (New York: Oxford University Press, 1984).

——, 'A Neutered Morality', *Third Way* September (1984), 27.

Omran, A. R., 'Epidemiologic Transition in the US: The Health Factor in Population Change', *Population Bulletin* 32/2 (1977), 1–42.

Orwell, George, *Nineteen Eighty-Four* (London: Secker and Warburg 1949).

Oxford English Dictionary, (Oxford: Oxford University Press, 2013).

Pandey, Subhash C., Adip Roy, Huaibo Zhang, and Tiejun Xu, 'Partial Deletion of the cAMP Response Element-Binding Protein Gene Promotes Alcohol-Drinking Behaviors', *The Journal of Neuroscience* 24/21 (2004), 5022–30.

Pannenberg, Wolfhart, *Jesus – God and Man*, Translated by Lewis L. Wilkins and Duane A. Priebe (London: SCM Press, 1968).

Pascual-Leone, A., A. Amedi, F. Fregni, and L. B. Merabet, 'The Plastic Human Brain Cortex', *Annual Review of Neuroscience* 28 (2005), 377–401.

Persaud, Raj, David Crossley, and Chris Freeman, 'Should Neurosurgery for Mental Disorder Be Allowed to Die Out?', *The British Journal of Psychiatry* 183/3 (2003), 195–6.

Persson, Ingmar, and Julian Savulescu, 'Getting Moral Enhancement Right: The Desirability of Moral Bioenhancement', *Bioethics* 27/3 (2013), 124–31.

——, 'The Perils of Cognitive Enhancement and the Urgent Imperative to Enhance the Moral Character of Humanity', *Journal of Applied Philosophy* 25/3 (2008), 162–77.

——, *Unfit for the Future: The Need for Moral Enhancement* (Oxford: Oxford University Press, 2012).

Peters, Ted, *Playing God? Genetic Determinism and Human Freedom* (New York: Routledge, 2003).

——, 'Resurrection of the Very Embodied Soul?', in Robert John Russell, Nancey Murphy, Theo C. Meyering and Michael A. Arbib, eds, *Neuroscience and the Person: Scientific Perspectives on Divine Action* (Indiana: University of Notre Dame Press, 1999), 305–26.

——, *Science, Theology and Ethics* (Aldershot: Ashgate, 2003).

——, 'The Soul of Trans-Humanism', *Dialog: A Journal of Theology* 44/4 (2005), 381–95.

——, *The Stem Cell Debate* (Minneapolis, MN: Fortress Press, 2007).

Peters, Ted, Karen Lebacqz, and Gaymon Bennett, *Sacred Cells? Why Christians Should Support Stem Cell Research* (Lanham, MD: Rowman & Littlefield Publishers, 2008).

Peterson, James C., *Changing Human Nature: Ecology, Ethics, Genes, and God* (Grand Rapids, MI: Wm. B. Eerdmans, 2010).

——, *Genetic Turning Points: The Ethics of Human Genetic Intervention* (Grand Rapids, MI: Wm. B. Eerdmans, 2001).

Pitman, R. K., K. M. Sanders, R. M. Zusman, A. R. Healy, F. Cheema, N. B. Lasko, L. Cahill, and S. P. Orr, 'Pilot Study of Secondary Prevention of Posttraumatic Stress Disorder with Propranolol', *Biological Psychiatry* 51/2 (2002), 189–92.

Plomin, R., and D. Daniels, 'Why Are Children in the Same Family So Different from One Another?', *International Journal of Epidemiology* 40/3 (2011), 563–82.

Plomin, Robert, Linzy Hill, Ian W. Craig, Peter McGuffin, Shaun Purcell, Pak Sham, David Lubinski, Lee A. Thompson, Paul J. Fisher, and Dragana Turic, 'A Genome-Wide Scan of 1842 DNA Markers for Allelic Associations with General Cognitive Ability: A Five-Stage Design Using DNA Pooling and Extreme Selected Groups', *Behavior Genetics* 31/6 (2001), 497–509.

Pohlmeyer, E. A., E. R. Oby, E. J. Perreault, S. A. Solla, K. L. Kilgore, R. F. Kirsch, and L. E. Miller, 'Toward the Restoration of Hand Use to a Paralyzed Monkey: Brain-Controlled Functional Electrical Stimulation of Forearm Muscles', *PLoS One* 4/6 (2009), e5924.

Pontifical Academy for Life, 'Final Communiqué on "the Dignity of Human Procreation and Reproductive Technologies. Anthropological and Ethical Aspects"', (February 2004) <http://www.vatican.va/roman_curia/pontifi-cal_academies/acdlife/documents/rc_pont-acd_life_doc_20040316_x-gen-assembly-final_en.html> accessed 3 April 2013.

Popper, K. R., and J. C. Eccles, *The Self and Its Brain: An Argument for Interactionism* (Berlin: Springer, 1977).

Powell, R., 'The Biomedical Enhancement of Moral Status', *Journal of Medical Ethics* 39/2 (2013), 65–6.

President's Council on Bioethics, *Beyond Therapy: Biotechnology and the Pursuit of Happiness* (Washington, DC, 2003).

——, *Human Cloning and Human Dignity: An Ethical Inquiry* (New York, NY: Public Affairs, 2002).

PRNewswire, 'Anatomist Dr. Gunther Von Hagens Reiterates His Mission of Public Health Education to Press Corps in Guben, Germany', (30 November 2006) <www.prnewswire.co.uk/cgi/news/release?id=185453> accessed 20 September 2010.

Radford, Tim, 'German Fury over Cloning', *The Guardian* (28 February 1997), 1.

Rakic, P., 'Neurogenesis in Adult Primate Neocortex: An Evaluation of the Evidence', *Nature Reviews: Neuroscience* 3/1 (2002), 65–71.

Ralston, A., and K. Shaw, 'Environment Controls Gene Expression: Sex Determination and the Onset of Genetic Disorders', *Nature Education* 1 (2008), 1.

Ramsey, Paul, 'Shall We "Reproduce"? I. The Medical Ethics of in Vitro Fertilization', *JAMA* 220/10 (1972), 1346–50.

——, 'Shall We "Reproduce"? II. Rejoinders and Future Forecast', *JAMA* 220/11 (1972), 1480–5.

Raser, J. M., and E. K. O'Shea, 'Noise in Gene Expression: Origins, Consequences, and Control', *Science* 309/5743 (2005), 2010–13.

Ratcliffe, Matthew, 'Neurotheology: A Science of What?', in Patrick McNamara, ed., *Where God and Science Meet: How Brain and Evolutionary Studies Alter Our Understanding of Religion* (Westport, CT: Praeger Publishers, 2006), 81–104.

Reichenbach, B. R., and V. E. Anderson, *On Behalf of God: A Christian Ethic for Biology* (Grand Rapids, MI: Eerdmans, 1995).

Retained Organs Commission, *A Consultation Document on Unclaimed and Unidentifiable Organs and Tissue, a Possible Regulatory Framework* (London: National Health Service, 2002).

Rice, George, Carol Anderson, Neil Risch, and George Ebers, 'Male Homosexuality: Absence of Linkage to Microsatellite Markers at Xq28', *Science* 284/5414 (1999), 665–7.

Robertson, J. A., 'Extending Preimplantation Genetic Diagnosis: The Ethical Debate. Ethical Issues in New Uses of Preimplantation Genetic Diagnosis', *Human Reproduction* 18/3 (2003), 465–71.

Rogers, Lois, 'Having Disabled Babies Will Be "Sin," Says Scientist', *Sunday Times* (4 July 1999).

Rogerson, John, *Theory and Practice in Old Testament Ethics* (London: T&T Clark International, 2004).

Rosen, Christine, 'Eugenics – Sacred and Profane', *New Atlantis* 2 (2003), 79–89.

Ryu, S. I., and K. V. Shenoy, 'Human Cortical Prostheses: Lost in Translation?', *Neurosurgical Focus* 27/1 (2009), E5.

Sacred Congregation for the Doctrine of the Faith, 'Declaration on Euthanasia', (1980) <http://www.vatican.va/roman_curia/congregations/cfaith/documents/rc_con_cfaith_doc_19800505_euthanasia_en.html> accessed 4 April 2013.

Sahakian, B., and S. Morein-Zamir, 'Professor's Little Helper', *Nature* 450 (2007), 1157–9.

Salkeld, L., 'Burly Rugby Player Has a Stroke after Freak Gym Accident … Wakes up Gay and Becomes a Hairdresser', *Daily Mail* (9 November 2011) <http://www.dailymail.co.uk/health/article-2058921/Chris-Birch-stroke-Rugby-player-wakes-gay-freak-gym-accident.html> accessed 22 June 2013.

Salzman, Mark, *Lying Awake* (New York, NY: Alfred A. Knopf, 2000).

Sandel, M. J., 'The Case against Perfection', *The Atlantic Monthly* 293 (2004), 51–62.

——, *The Case against Perfection: Ethics in the Age of Genetic Engineering* (Cambridge, MA: Belknap Press, 2007).

Sapolski, R., 'Chaos and Reductionism', Stanford University (19 May 2010) <http://www.youtube.com/watch?v=_njf8jwEGRo> accessed 18 June 2013.

Saver, Jeffrey L., and John Rabin, 'The Neural Substrates of Religious Experience', *The Journal of Neuropsychiatry and Clinical Neurosciences* 9 (1997), 498–510.

Savulescu, J., 'Genetic Interventions and the Ethics of Enhancement of Human Beings', in Bonnie Steinbock, ed., *The Oxford Handbook of Bioethics* (Oxford: Oxford University Press, 2007), 516–35.

Savulescu, J., T. Douglas, and I. Persson, 'Autonomy and the Ethics of Behavioural Modification', in Akira Akabayashi, ed., *The Future of Bioethics: International Dialogues* (Oxford: Oxford University Press, 2014), 91–112.

Savulescu, J., B. Foddy, and M. Clayton, 'Why We Should Allow Performance Enhancing Drugs in Sport', *British Journal of Sports Medicine* 38/6 (2004), 666–70.

Savulescu, J., and A. Sandberg, 'Neuroenhancement of Love and Marriage: The Chemicals between Us', *Neuroethics* 1 (2008), 31–44.

Schjodt, U., H. Stodkilde-Jorgensen, A. W. Geertz, and A. Roepstorff, 'Rewarding Prayers', *Neuroscience Letters* 443/3 (2008), 165–8.

Schlesselman, J. J., 'How Does One Assess the Risk of Abnormalities from Human in Vitro Fertilisation?', *American Journal of Obstetrics and Gynecology* 135 (1979), 135–48.

Schulte-Sasse, Linda, 'Advise and Consent: On the Americanization of Body Worlds', *BioSocieties* 1 (2006), 369–84.

Segal, David, 'This Man Is Not a Cyborg. Yet', *New York Times* (1 June 2013) <http://www.nytimes.com/2013/06/02/business/dmitry-itskov-and-the-avatar-quest.html?_r=0&adxnnl=1&pagewanted=all&adxnnlx=1372069074-ckc-c4DUQvoAkkAIrxBOoSw> accessed 24 June 2013.

Shannon, T. A., and J. J. Walter, *The New Genetic Medicine: Theological and Ethical Reflections* (Lanham, MD: Rowman & Littlefield, 2003).

Shostak, S., *Becoming Immortal: Combining Cloning and Stem-Cell Therapy* (Albany, NY: State University of New York Press, 2002).

Silver, L. M., *Remaking Eden: Cloning and Beyond in a Brave New World* (New York, NY: Aven Books, 1997).

Singer, Peter, and Agata Sagan, 'Are We Ready for a "Morality Pill"?', *New York Times* (28 January 2012) <http://opinionator.blogs.nytimes.com/2012/01/28/are-we-ready-for-a-morality-pill/> accessed 23 June 2013.

Skulstad, Kenji, 'Body Worlds Draws Large Crowds – and Controversy', *Canadianchristianity.com* (4 July 2006) <www.canadianchristianity.com/cgi-bin/bc.cgi?bc/bccn/1106/18body> accessed 12 October 2007.

Smith, Caspar Llewellyn, 'Aubrey De Grey: We Don't Have to Get Sick as We Get Older', *The Observer* (1 August 2010).

Song, Robert, 'Genetic Manipulation and the Resurrection Body', in King-Tak Ip, ed., *The Bioethics of Regenerative Medicine* (Netherlands: Springer, 2009), 27–45.

——, 'To Be Willing to Kill What for All One Knows Is a Person Is to Be Willing to Kill a Person', in B. Waters and R. Cole-Turner, eds, *God and the Embryo: Religious Voices on Stem Cells and Cloning* (Washington, DC: Georgetown University Press, 2003), 98–107.

Southern Baptist Convention, 'Resolution on Genetic Technology and Cloning', (June 1997) <http://www.sbc.net/resolutions/amResolution.asp?ID=571> accessed 29 May 2013.

Sparrow, R. J., 'The Perils of Post-Persons', *Journal of Medical Ethics* 39/2 (2013), 80–1.

Spezio, Michael L., 'Brain and Machine: Minding the Transhuman Future', *Dialog: A Journal of Theology* 44/4 (2005), 375–80.

Stagg, Charlotte J., and Michael A. Nitsche, 'Physiological Basis of Transcranial Direct Current Stimulation', *The Neuroscientist* 17/1 (2011), 37–53.

Statistical Directorate, 'Statistical Bulletin, Life Expectancy 2007–2009', Welsh Assembly Government (24 November 2010) <http://wales.gov.uk/docs/stat istics/2010/101124sb942010en.pdf> accessed 26 June 2013.

Steele, J. Douglas, David Christmas, M. Sam Eljamel, and Keith Matthews, 'Anterior Cingulotomy for Major Depression: Clinical Outcome and Relationship to Lesion Characteristics', *Biological Psychiatry* 63/7 (2008), 670–7.

Steinberg, L., 'The Influence of Neuroscience on US Supreme Court Decisions About Adolescents' Criminal Culpability', *Nature Reviews: Neuroscience* 14/7 (2013), 513–18.

Stern, Megan, 'Dystopian Anxieties Versus Utopian Ideals: Medicine from *Frankenstein* to the *Visible Human Project* and *Body Worlds*', *Science as Culture* 15/1 (2006), 61–84.

——, 'Shiny Happy People: "Body Worlds" and the Commodification of Health', *Radical Philosophy* 118 (2003), 2–6.

Stiles, J., 'Brain Development and the Nature Versus Nurture Debate', *Progress in Brain Research* 189 (2011), 3–22.

Stock, G., *Choosing Our Children's Genes: Redesigning Humans* (London: Profile Books, 2002).

——, *Redesigning Humans: Our Inevitable Genetic Future* (New York, NY: Houghton Mifflin Company, 2002).

Streatfeild, Dominic, *Brainwash: The Secret History of Mind Control* (London: Hodder & Stoughton, 2006).

Strom, C. M., N. Ginsberg, M. Applebaum, N. Bozorgi, M. White, M. Caffarelli, and Y. Verlinsky, 'Analyses of 95 First-Trimester Spontaneous Abortions by Chorionic Villus Sampling and Karyotype', *Journal of Assisted Reproduction and Genetics* 9 (1992), 458–61.

Swinton, J., and Brock B., eds, *Theology, Disability and the New Genetics: Why Science Needs the Church* (London: T&T Clark, 2007).

Szreter, S., and G. Mooney, 'Urbanisation, Mortality, and the Standard of Living Debate; New Estimates of the Expectation of Life at Birth in Nineteenth-Century British Cities', *The Economic History Review* 50 (1998), 84–112.

Tachibana, Masahito, Paula Amato, Michelle Sparman, Nuria Marti Gutierrez, Rebecca Tippner-Hedges, Hong Ma, Eunju Kang, Alimujiang Fulati, Hyo-Sang Lee, and Hathaitip Sritanaudomchai, 'Human Embryonic Stem Cells Derived by Somatic Cell Nuclear Transfer', *Cell* 153/6 (2013), 1228–38.

Tamburrini, Guglielmo, 'Brain to Computer Communication: Ethical Perspectives on Interaction Models', *Neuroethics* 2 (2009), 137–49.

Tancredi, L. R., and J. D. Brodie, 'The Brain and Behavior: Limitations in the Legal Use of Functional Magnetic Resonance Imaging', *American Journal of Law & Medicine* 33/2–3 (2007), 271–94.

Tang, Thomas, and Adam H. Balen, 'Use of Metformin for Women with Polycystic Ovary Syndrome', *Human Reproduction Update* 19/1 (2013), 1.

Tang, Ya-Ping, Eiji Shimizu, Gilles R. Dube, Claire Rampon, Geoffrey A. Kerchner, Min Zhuo, Guosong Liu, and Joe Z. Tsien, 'Genetic Enhancement of Learning and Memory in Mice', *Nature* 401/6748 (1999), 63–9.

Taylor, Kathleen, *Brainwashing: The Science of Thought Control* (Oxford: Oxford University Press, 2006).

Taylor, Philippa, *For What It Is Worth: The Status of the Human Embryo* (London: CARE/CBPP, 2002).

Thomas, Dylan, *Collected Poems 1934–1952* (London: Dent and Sons, 1952).

Thomson, J. A., J. Itskovitz-Eldor, S. S. Shapiro, M. A. Waknitz, J. J. Swiergiel, V. S. Marshall, and J. M. Jones, 'Embryonic Stem Cell Lines Derived from Human Blastocysts', *Science* 282/5391 (1998), 1145–7.

Tirosh-Samuelson, H., and K. L. Mossman, eds, *Building Better Humans? Refocusing the Debate on Transhumanism* (Frankfurt am Main: Peter Lang, 2012).

Todd, Rebecca M., and Adam K. Anderson, 'Six Degrees of Separation: The Amygdala Regulates Social Behaviour and Perception', *Nature Neuroscience* 12/10 (2009), 1217–18.

Torrance, Thomas F., *Test-Tube Babies: Morals – Science – and the Law* (Edinburgh: Scottish Academic Press, 1984).

Tost, Heike, and Andreas Meyer-Lindenberg, 'I Fear for You: A Role for Serotonin in Moral Behavior', *Proceedings of the National Academy of Sciences* 107/40 (2010), 17071–2.

Towns, C. R., and D. G. Jones, 'Stem Cells: Public Policy and Ethics', in M. Ruse and C. A. Pynes, eds, *The Stem Cell Controversy: Debating the Issues* (New York, NY: Prometheus Books, 2004), 329–41.

——, 'Stem Cells: Public Policy and Ethics', *NZ Bioethics Journal* 5 (2004), 22–8.

Turner, D. C., T. W. Robbins, L. Clark, A. R. Aron, J. Dowson, and B. J. Sahakian, 'Cognitive Enhancing Effects of Modafinil in Healthy Volunteers', *Psychopharmacology* 165/3 (2003), 260–9.

Utz, K. S., V. Dimova, K. Oppenländer, and G. Kerkhoff, 'Electrified Minds: Transcranial Direct Current Stimulation (tDCS) and Galvanic Vestibular Stimulation (GVS) as Methods of Non-Invasive Brain Stimulation in Neuropsychology – a Review of Current Data and Future Implications', *Neuropsychologia* 48/10 (2010), 2789–810.

Vaiva, G., F. Ducrocq, K. Jezequel, B. Averland, P. Lestavel, A. Brunet, and C. R. Marmar, 'Immediate Treatment with Propranolol Decreases Posttraumatic Stress Disorder Two Months after Trauma', *Biological Psychiatry* 54/9 (2003), 947–9.

Valenstein, Elliot S., *Great and Desperate Cures: The Rise and Decline of Psychosurgery and Other Radical Treatments* (New York, NY: Basic Books, 1986).

van Ommen, G. J., E. Bakker, and J. T. den Dunnen, 'The Human Genome Project and the Future of Diagnostics, Treatment, and Prevention', *Lancet* 354, Suppl. 1 (1999), S5–10.

Vastag, Brian, 'Poised to Challenge Need for Sleep, "Wakefulness Enhancer" Rouses Concerns', *JAMA: The Journal of the American Medical Association* 291/2 (2004), 167–70.

Velliste, M., S. Perel, M. C. Spalding, A. S. Whitford, and A. B. Schwartz, 'Cortical Control of a Prosthetic Arm for Self-Feeding', *Nature* 453/7198 (2008), 1098–101.

Verhey, Allen, *The Christian Art of Dying: Learning from Jesus* (Grand Rapids, MI: Wm. B. Eerdmans Publishing, 2011).

——, '"Playing God" and Invoking a Perspective', *Journal of Medicine and Philosophy* 20/4 (1995), 347–64.

——, *Reading the Bible in the Strange World of Medicine* (Grand Rapids, MI: W. B. Eerdmans Publishing, 2003).

——, *Remembering Jesus: Christian Community, Scripture, and the Moral Life* (Grand Rapids, MI: Wm. B. Eerdmans Publishing, 2002).

——, 'What Makes Christian Bioethics Christian? Bible, Story, and Communal Discernment', *Christian Bioethics* 11/3 (2005), 297–315.

Vincent, J. A., 'Ageing Contested: Anti-Ageing Science and the Cultural Construction of Old Age', *Sociology* 40/4 (2006), 681–98.

von Hagens, Gunther, 'Anatomy and Plastination', in Gunther von Hagens and Angelina Whalley, eds, *Anatomy Art – Fascination beneath the Surface* (Heidelberg: Institute for Plastination, 2000), 11–38.

Voon, V., C. Kubu, P. Krack, J. L. Houeto, and A. I. Troster, 'Deep Brain Stimulation: Neuropsychological and Neuropsychiatric Issues', *Movement Disorders* 21/Suppl 14 (2006), S305–27.

Vorster, J. M., 'A Christian Ethical Perspective on the Moral Status of the Human Embryo', *Acta Theologica* 31/1 (2011), 189–204.

Walsh, Fergus, '10,000 NHS Patients "to Have Genes Mapped"', *BBC News* (21 June 2010) <http://www.bbc.co.uk/news/10367883> accessed 5 April 2013.

Walter, Tony, 'Plastination for Display: A New Way to Dispose of the Dead', *Journal of the Royal Anthropological Institute* 10 (2004), 603–27.

Warnock, Mary, *Report of the Committee of Inquiry into Human Fertilization and Embryology* (London: HMSO, 1984).

Waters, B., 'Does the Human Embryo Have a Moral Status?', in B. Waters and R. Cole-Turner, eds, *God and the Embryo: Religious Voices on Stem Cells and Cloning* (Washington, DC: Georgetown University Press, 2003), 67–76.

Waters, B., and R. Cole-Turner, eds, *God and the Embryo: Religious Voices on Stem Cells and Cloning* (Washington, DC: Georgetown University Press, 2003).

Wield, Cathy, *Life after Darkness: A Doctor's Journey through Severe Depression* (Oxford: Radcliffe, 2006).

——, *A Thorn in My Mind: Mental Illness, Stigma and the Church* (Watford: Instant Apostle, 2012).

Wilmut, I., A. E. Schnieke, J. McWhir, A. J. Kind, and K. H. Campbell, 'Viable Offspring Derived from Fetal and Adult Mammalian Cells', *Nature* 385/6619 (1997), 810–13.

Wolf, G. C., and E. O. Horger, 'Indications for Examination of Spontaneous Abortion Specimens: A Reassessment', *American Journal of Obstetrics and Gynecology* 173 (1995), 1364–8.

Wong, A. H., I. I. Gottesman, and A. Petronis, 'Phenotypic Differences in Genetically Identical Organisms: The Epigenetic Perspective', *Human Molecular Genetics* 14/Suppl 1 (2005), R11–18.

World Health Organization, 'Life Expectancy by Country', <http://apps.who.int/gho/data/node.main.688?lang=en> accessed 26 June 2013.

Wright, N. T., *The Resurrection of the Son of God: Christian Origins and the Question of God* (London: SPCK, 2003).

——, *Surprised by Hope: Rethinking Heaven, the Resurrection, and the Mission of the Church* (New York, NY: HarperOne, 2008).

Wyatt, John, *Matters of Life and Death*, 2nd edn (Nottingham: Inter-Varsity Press, 2009).

Yesavage, J. A., M. S. Mumenthaler, J. L. Taylor, L. Friedman, R. O'Hara, J. Sheikh, J. Tinklenberg, and P. J. Whitehouse, 'Donepezil and Flight Simulator Performance: Effects on Retention of Complex Skills', *Neurology* 59/1 (2002), 123–5.

Yong, A., *Theology and Down Syndrome: Reimagining Disability in Late Modernity* (Waco, TX: Baylor University Press, 2007).

Zins, J. E., and A. Moreira-Gonzalez, 'Cosmetic Procedures for the Aging Face', *Clinics in Geriatric Medicine* 22/3 (2006), 709–28.

Index of General Terms

Index of Names

New International Studies in Applied Ethics

SERIES EDITORS
Professor R. John Elford and Professor Simon Robinson,
Leeds Metropolitan University

New International Studies in Applied Ethics is a series based at Leeds Metropolitan University and associated with Virginia Theological Seminary. The series examines the ethical implications of selected areas of public life and concern. Subjects considered include, but are not limited to, medicine, peace studies, international sport and higher education.

The series aims to publish volumes which are clearly written with a general academic readership in mind. Individual volumes may also be useful to those confronted with the issues discussed in their daily lives. A consistent emphasis is on recent developments in the subjects discussed and this is achieved by publishing volumes by writers who are foremost in their fields, as well as those with emerging reputations. Both secular and religious ethical views may be discussed as appropriate. No point of view is considered off-limits and controversy is not avoided.

The series includes both edited volumes and single-authored monographs. Submissions are welcome from all scholars in the field and should be addressed to either the series editors or the publisher.

Vol. 1   R. John Elford and D. Gareth Jones (eds)
          A Tangled Web: Medicine and Theology in Dialogue
          288 pages. 2009.
          ISBN: 978-3-03911-541-9

Vol. 2   D. Gareth Jones and R. John Elford (eds)
          A Glass Darkly: Medicine and Theology in Further Dialogue
          254 pages. 2010.
          ISBN: 978-3-03911-936-3